Site Reliability Engineering (SRE) Principles: Ensuring Uptime, Scalability, and Efficiency

Gopikrishna Maddali
Swapnil J. Wawge

Made with ♥ on the Notion Press Platform

www.notionpress.com

Dedications

To every engineer who has turned panic into process and outages into lessons.

Contents

Preface

Site Reliability Engineering (SRE) Principles: Ensuring Uptime, Scalability, and Efficiency

In an era where digital experiences shape business success, system reliability is no longer a luxury—it's a necessity. Whether it's a global e-commerce platform, a real-time financial service, or a healthcare application delivering critical information, users expect near-instant access and uninterrupted service. The need to build systems that are not only fast and responsive but also resilient and scalable has given rise to a transformative discipline: Site Reliability Engineering.

This book is born out of the growing demand for a deeper understanding of how complex software systems can be operated efficiently and reliably at scale. It distills the core principles, practices, and philosophies of SRE into a comprehensive guide for engineers, architects, managers, and anyone committed to delivering high-availability services. Drawing inspiration from industry pioneers like Google and expanding upon practical lessons from a variety of environments, this work explores the fusion of software engineering with IT operations to address modern infrastructure challenges.

Throughout these chapters, we will delve into topics including service-level objectives (SLOs), incident response, automation, observability, capacity planning, and the human side of operational excellence. Each principle is grounded in real-world examples and actionable insights designed to help teams build a culture of reliability, drive efficiency, and maintain agility in rapidly evolving environments.

This book is not just for those already practicing SRE—it is also for those aspiring to adopt its mindset, principles, and tools to improve their systems and teams. Whether you are starting your SRE journey or looking to deepen your expertise, we hope this guide becomes a valuable companion in your pursuit of sustainable, scalable, and efficient reliability.

Acknowledgment

Writing Site Reliability Engineering (SRE) Principles: Ensuring Uptime, Scalability, and Efficiency has been a journey shaped by the collective wisdom, support, and inspiration of many remarkable individuals.

First and foremost, we extend our deepest gratitude to the global SRE community—engineers, architects, and thought leaders, whose shared knowledge, open-source contributions, and insightful discussions have laid the foundation for this book. Your passion for reliability, automation, and continuous improvement continues to push the boundaries of what's possible in system design and operations.

A heartfelt thank you goes to my mentors and colleagues—those who challenged my assumptions, offered guidance during late-night incident reviews, and showed me the value of both rigor and empathy in engineering. Your stories and experiences are woven into the fabric of this book.

To my technical reviewers and peers who provided invaluable feedback, thank you for your sharp eyes, critical insights, and unwavering encouragement. Your contributions helped shape this work into something stronger and more relevant.

To my friends and family, your patience and understanding during the many hours spent writing and revising did not go unnoticed. Your support made it possible for me to focus, reflect, and persist.

Lastly, to the readers of this book—whether you're an SRE veteran, a curious developer, or a team leader looking to bring reliability thinking into your organization, thank you for your interest and trust. We hope these pages serve you well on your path toward building systems that are not only robust and scalable but also humane and maintainable.

CHAPTER 01

Introduction to Site Reliability Engineering (SRE)

1.1 Defining Site Reliability Engineering

Site reliability engineering (SRE), as an engineering discipline, has emerged from the requirements for a more sophisticated use of modern software systems to become more complex and the need to scale. In turn, it integrates principles regarding computer science, software engineering, systems architecture and operational management to provide the basis for a unifying framework for maintaining large scale distributed systems in terms of reliability, availability, scalability, and efficiency (Pethuru Raj Chelliah, Shreyash Naithani, 2018).

SRE was first pioneered by Google in the early 2000s as a means to understand how to automatically operate dynamic, elastic, and cloud native infrastructures, they were simply no longer handled particularly effectively using traditional system administration techniques. Rather than react to failures, SRE brings an automation first, proactively focus on preventing failures and reducing the impact of failures when they occur. It defines operations as a software problem that can be measured, managed and improved continuously through the same tools and methodologies learned in software development.

Site Reliability Engineering is based on the idea of defining and keeping Service Level Indicators (SLIs), Service Level Objectives (SLOs), Service Level Agreements (SLAs) at the core. These classes of construct are utilized as formal benchmarking ladders to measure the performance, uptime and customer satisfaction of a system. SLIs measure aspects of service behaviour like latency, availability, error rates, throughputs etc. SLOs define the targets for these indicators and SLAs define contractual expectations from external stakeholders. SRE teams keep a continuous eye on these metrics to measure system health and can do so in real business terms to prioritize engineering tasks.

One of the keys of SRE is automation, to the point where automated, not manual, is considered 'better.' That is to say, SRE deems the automation of tasks that don't contribute to lasting value to be a good thing. To mitigate the effect of human intervention, improve consistency for deployment, while maintaining velocity, SRE practitioners develop and deploy automated tools and scripts in order to reduce human touch points in the process. This principle is in line with the overall goals of DevOps, but gives a special emphasis on the mathematical rigor and the service-level accountability.

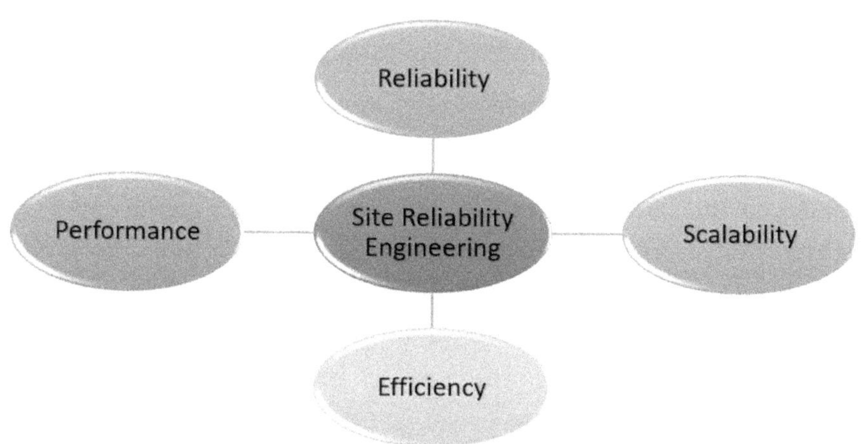

Figure 1.1: Key Aspects of SRE: A Concept Map.

Source: (Kottisi, 2023)

Incident management, the overall management of an incident and postmortem analysis is another central pillar of SRE paradigm. Culture is so important in SRE because SRE promotes a blameless culture: in such a culture, system failures are seen as opportunities for learning and for systemic improvement. The post incident reviews are done to find out the root causes, the lessons learnt and the measures to be taken or implemented. The culture of resilience engineering considers the system design patterns such as redundancy, graceful degradation and failover mechanisms to increase fault tolerance.

In addition, Site Reliability Engineering joins the gap between development and operations, serving as a keystone, that there is no point in speeding up the engineering velocity faster than the system reliability. This is because by working very closely with development teams, SREs help design and code for systems that are easier to observe, test and maintain. To monitor system health in real time, they rely on sophisticated observability frameworks and mirror diagnose using advanced anomaly detection and trend analysis tools.

In sum up, Site Reliability Engineering is not just a role nor a set of tools, but a new attitude and methodological framework that allows to put scientific rigor on the operation of complex systems. It reorients the lines between software engineering and IT operations, restructuring the ways one deliveries and sustains services at scale. SRE principles like reliability as a feature, error budgets, continuous feedback, automation, and proactive monitoring allow the organization to innovate fast and retain the robustness and trustworthiness of their digital services.

1.1.1 The Origin of SRE: Google's Influence

Site Reliability Engineering (SRE) was invented by Google as the answer to its original operational problems in the early 2000s as it grew to become the biggest and most complicated digital service provider in the world. For traditional operations teams, growing to 'unprecedented' level of scale, user demand and system complexity was impossible to scale. In Google's engineers, they saw that in order to sustain high availability and performance at global scale, a new paradigm was needed; one that would not only exist beyond manual administration and reactive firefighting. To solve this, Google reinvented the way systems should be kept and used and created a specialized discipline that relied on engineering talent to exclusively work on the health and robustness of production systems. In this case, it was the formal birth of SRE, not as an offshoot of IT operations, but as a new branch of engineering with its own methodologies, responsibilities, and cultural ethos (Niall Richard Murphy, Chris Jones, 2016a).

In contrast to most other organizations at the time, Google made a strategic choice to hire SRE teams primarily with software engineers, rather than traditional system administrators. This was a fundamental change in philosophy: instead of being reliant on human intervention to keep infrastructure reliable, SRE wanted to build self-healing, scalable infrastructure using tools, code and automation. This structure was an innovation that allowed SRE teams to be deeply embedded in the development lifecycle and to collaborate closely with product teams in early design discussions, and bring reliability considerations into early discussions. Gone was the need to wait for the network division to assess the impact of a change to system reliability; SREs were granted the ability to say no to changes with too much risk, and to hold themselves accountable for system uptime. The tension between these two led to an engineering rigor in reliability planning and proactive reliability work, which was shared between development and operations.

Moreover, many of the modern reliability strategies that have become standard in the tech industry were based on Google's internal practices. One such example is structured incident response process Google took to enable engineers to act quickly when certain incidents occurred — such as runbooks and automation scripts for posts. The more important one was to establish a culture of blameless postmortem, a culture in which failures were to be perceived as opportunities to discover and fix latent weaknesses in the system. Sharing these learnings openly meant that recurrence was prevented and the system had higher systemic resilience. These practices grew up around themselves and evolve into codified process, which eventually suffice to share with the public via books, talks, and open-source tools, and now the Google isn't just the epicenter of SRE core but a global evangelist of reliability engineering. Modern digital enterprises can still feel the effects of the company's early experiences when it comes to designing for reliability, thinking about risk, and scaling operational excellence in a distributed architecture.

1.1.2 The Core Philosophy of SRE

Site Reliability Engineering (SRE) is the shifting of how modern organizations think about managing large scale distributed systems, and at the center of this philosophy is a peculiar and pragmatic set of ideas that redefine what should drive how organizations manage modern complex system. In contrast to the traditional operations model, which heavily depends on the manual intervention and secretive responsibilities, the SRE philosophy propagates that the software and automation are the equipment that primarily proves reliability productively. Essentially, reliability is not an afterthought, a reactive quality, but rather, its core belief, to be a fundamental aspect of system design and service delivered. It also encourages these proactive engineering practices of anticipative failure modes, build mechanisms of graceful degradation and recovery strategies into the architecture as an inherent part of it. As such, reliability is placed on the same footing as feature development and every design decision is considered with respect to how it affects availability and user experience (Ukis, 2022).

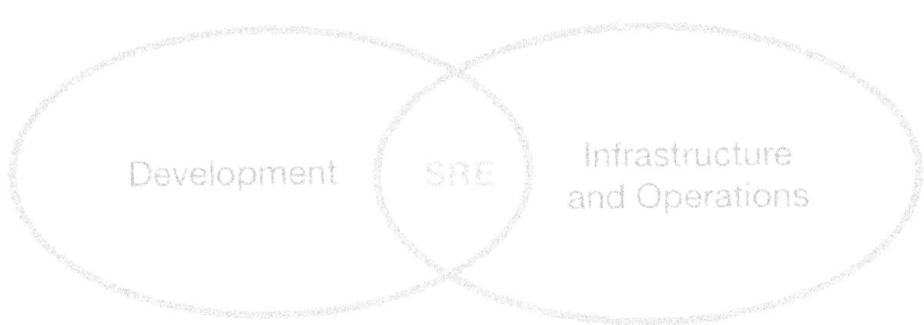

Figure 1.2: Venn Diagram of Site Reliability Engineering.

Source: (Gunderson, 2021)

A second essential aspect of the SRE philosophy is the measurement of reliability based on measurable objectives and accepted levels of tolerance in risk. Rather than shooting for a level of absolute perfection, error budgeting is the approach to being ok with some failure, and in fact, even error budgets are part of the SRE philosophy. Service Level Objectives (SLOs) help define what targets you have for your availability, and then you track deviations in Service Level Indicators (SLIs) to manage the trade between system stability and development velocity through data. This philosophical shift enables teams to make some informed decisions about when they should focus on reliability work instead of doing new things, cultivating an accountable and disciplined sense of operational reality. It calculates subjective discussions into objective evaluation which makes technical performance a representation of business priorities (Smith, 2021).

The last foundational piece of the SRE philosophy is the cultural aspect of being resilient, learning, and collaborating. SRE practices instill a blameless culture whereby engineers are not afraid of identifying and clearing the root cause of the incident without being held responsible. Not only does this make psychological safety better, but it also reduces the speed that it takes to learn as an organization by converting failures into institutional knowledge. Incidents are an opportunity for systemic improvement in SRE rather than individual fault leading to continuous evolution of both the tools and processes. Secondly, in addition, the SRE teams act as a bridge between software development and operations, with a very close collaboration and shared ownership of the system reliability. The foundation of that is a holistic philosophy: engineering based, metric guided and culture empowered, which is the essence of what makes SRE a powerful discipline for managing modern and complex systems.

1.2 The Evolution of SRE: From DevOps to Modern Reliability Practices

Site Reliability Engineering (SRE) is an evolutionary journey by organizations as they learn to create reliable systems that deliver a high quality of service within an operationally excellent environment in this digitized age. Due to the increasing level of complexity of software systems, their global distribution and, for ever more mission critical systems, the expectations for the availability and performance have risen. To meet these expectations the paradigm must go beyond traditional administration or cultural philosophies, this calls for an engineering-oriented discipline of reliability as a core system attribute (Singpurwalla, 2007).

At first, the industry tackled the gap between operation and development by means of process improvements and automation techniques. Yet, this stage of progress was not enough for addressing the ongoing problems of observability, capacity management, incident response and performance tuning at scale. The need to integrate system reliability directly into the software lifecycle was recognized by organizations and they began to realize the importance of having a structured framework that enabled engineering of fault tolerant systems from the ground up. Therefore, this realization led to the maturity of SRE as a standalone discipline that is much beyond automation or process agility.

Modern SRE is a sophisticated practice based on continuous system improvement, proactive risk mitigation and strategy in reliability planning. It has methods like FMEA, Dependency mapping, Proactive Chaos injection to find and kill the single point of failure. Additionally, SRE is well integrated with agile development cycles and allows teams to push the frequent deployment with stable system. Taking SRE to its pinnacle, this is the use of AI at anomaly detection, root causes analysis and automated response systems as examples, showing how SRE has become a future proof way of your engineering precision with your operational awareness. In the context of an ever more dynamic world of technology landscapes, SRE is continuously developing and changing the way a business creates, operates and maintains the digital infrastructures of tomorrow.

This is an important aspect of the evolution because it also represents a cultural sea change in the level of importance attached to operational expertise. It has stopped being affiliated to support or reactive function, now instead serving as a strategic enabler of business continuity, customer trust and engineering efficiency. Modern SRE practices, through the shared responsibility of reliability within cross functional teams embeds reliability to be a shared responsibility, allowing for the culture of collaboration, learning, and accountability. Such a shift creates more robust systems, as well as quicker tickets and a more proactive 'ahead of the curve' mindset that is always one step ahead of risks, making SRE a key piece of the architecture of being in a modern digital business.

1.2.1 The Shift from Traditional IT Operations

A very pronounced shift in both thinking and doing has taken place within the technology landscape characterized as the transition from traditional IT operations to modern reliability centered practice. Most of the traditional IT operations were based on a centralized, hierarchical structure wherein the operation teams were primarily responsible for the deployment, maintenance, and stability of the production systems. These teams were in control of infrastructure, manual configuration, periodic software updates and incident resolution. Most of the time, they relied on their predefined runbooks and ticketing systems, and they had very strict control over change management to prevent service disruption. Since, system updates were infrequent and incident response was reactive based on human intervention after an outage or failure (Duderstadt, 2010).

However, this model is perfectly fitting to the world of a monolithic application and an on-premises data center of an earlier era, where changes to the production environments were rare and predictable. But requirements for IT operations changed drastically with the arrival of agile development methodologies, cloud computing and the path of microservices architecture. Today, applications are being deployed many times per day in highly distributed and dynamic infrastructures. The customer's expectation for real time availability, low latency and continuous delivery has significantly escalated. In this new context, I found that the traditional approach had serious shortcomings that included lack of scalability, speed, adaptability and efficient execution of the interdependencies in a complex system.

When it was realized that infrastructure was evolving and required adaptation to keep pace with the fast-changing software it was built to support, the traditional model was replaced. To solve this, the modern reliability engineering, especially exemplified with Site Reliability Engineering (SRE) was born to leverage the software engineering principles into the realm of operations. The crucial point was that this change focused more on automation than manual work, active system monitoring and more passive alerting, and service reliability measured in performance rather than the presumed stability of the system. The development and deployment process for a modern application goes through infrastructure as code (IaC), continuous integration and continuous delivery (CI/CD) pipelines, real time observability, and incident management automation both enabled operations to do infrastructure in scale, with consistency and repeatability (Ukis, 2022).

In addition, this shift has also completely changed the roles and responsibilities of the people within the software delivery teams. The days of operation and development glossing over one another, with the operations people working under one piece of equipment control and the development people working under another, are over – operations and development were integrated into collaboratively responsible units sharing the performance, reliability and user experience of the system. Blameless postmortems, error budgets and service level objectives (SLOs) are all modern practices that encourage reliability to be quantifiable, designable rather than an afterthought. These principles have enabled organizations to move from fear driven stable culture to innovative, and resilient culture (Turner, 2021).

In fact, the transition away from classic IT operations to practices of modern reliability is not a technological update, but a cultural reengineering. It is a manifestation of the engineering systems towards scalability, observability, maintainability, robustness, and performability that complements performance and robustness. It is this paradigm change that puts in place the ability to deliver highly available digital services at a global scale by shifting from the posture of an organization that is reactive, maintenance focused, to that of an organization which is proactive, contributing strategically to the success of the business in the digital age.

1.2.2 Differences Between DevOps and SRE

DevOps and Site Reliability Engineering (SRE) are both trying to solve this gap between software development and IT ops, but are so different on what they are, how they're implemented, tooling, and the organizational structure. Both disciplines were born to solve the same problems (speeding the delivery of software, while maintaining reliability) but they use different tools and reactions. The organizations that wish to adopt right reliability and delivery strategies in their environments need to understand their differences.

It is a cultural and organizational philosophy that unites the developers with the operations team, literally called DevOps. It also describes CI/CD, its role of reducing development cycles and increasing deployment frequency and continuous integration, and provides the example of CI/CD from the perspective of automation. DevOps encourages some of the infra as code, monitoring and feedback loops practices to ensure that releases of software become of higher quality. In contrast to prescriptiveness on roles, metrics and how to implement it, it does not mandate, but rather a wide tool set of principles to be adapted to any organizational need (Coupland, 2021).

But on the contrary, SRE is a particular engineering discipline with structured and measurable approach to reliability. It also provides concrete objects in the form of Service Level Indicators (SLIs), Service Level Objectives (SLOs) and error budgets to measure, monitor, and control service reliability. The software engineers involved in reliability, scalability, performance and automation of the system are called as SRE teams. In contrast, SRE is an engineering solution to operational problems like automating toil, writing code that removes repetitive tasks, building fault tolerant systems up front, and so on.

The second main difference is how they deal with incidents and risk. Both DevOps teams and SRE have processes in place for general monitoring and incident response, but SRE relies on blameless postmortems and quantitative error budget policies to determine optimality between reliability and feature delivery. It has more formal means of risk tolerance and capacity planning, and service health is always monitored and improved (Reason, 2016).

From an organizational structure, DevOps is a team ownership and the entire engineering organization is to be responsible for DevOps and teams take up DevOps practices in a team context. Unlike SRE, an SRE team is often part of an organization with its own product and development teams and they work with them to improve service performance and enforce reliability standards.

Table 1.1: Key Differences Between DevOps and SRE.

Aspect	**DevOps**	**SRE (Site Reliability Engineering)**
Definition	A cultural and procedural philosophy promoting collaboration between development and operations teams.	An engineering discipline focused on building and maintaining reliable, scalable systems.
Primary Focus	Speed, agility, automation, and continuous delivery.	Reliability, availability, system performance, and scalability.
Approach to Reliability	Implicitly handled through DevOps practices like CI/CD, monitoring, and testing.	Explicitly managed using SLIs, SLOs, and error budgets.

Aspect	**DevOps**	**SRE (Site Reliability Engineering)**
Team Structure	Shared responsibility across dev and ops teams.	Usually a dedicated team of engineers with reliability-specific roles.
Metrics and Measurement	No strict framework; varies by organization.	Uses formal metrics (SLIs, SLOs, error budgets) to measure and enforce reliability.
Incident Management	Reactive handling using monitoring and alerts.	Proactive handling through automated tools, blameless postmortems, and structured analysis.
Tooling	Emphasizes CI/CD tools, version control, config management, and containers.	Includes reliability-focused tools for chaos testing, observability, and capacity planning.
Automation	Strongly encouraged, but not systematically enforced.	Essential to eliminate toil and ensure scalable operations.
Cultural Emphasis	Collaboration, shared ownership, and communication.	Engineering rigor, risk tolerance, and operational excellence.
Origins	Grassroots, industry-wide movement.	Institutionalized practice originating from engineering disciplines.

Source: (Self-generated)

In summary, even then, SRE has designed and inculcated a pattern of system reliability at spec and operation level using DevOps' culture and procedural core to bring more integration of development and operations. In most organizations, the maximum value is achieved by combining the two: DevOps to foster collaboration and velocity, and

SRE to keep the velocity from deteriorating the service stability and user trust.

1.2.3 Enhancing DevOps Methodologies Through SRE

Site Reliability Engineering is a transformative leap in modern software engineering and in areas that are significantly important in software engineering, for example, system deployment, operational resilience and lifecycle management. While DevOps is a cultural and procedural framework, its purpose is to break down the traditional barriers between software development and IT operations by promoting collaboration, continuous integration, and fast development pipelines. Still, despite being a service-oriented approach to solve agile and automated issues, DevOps was missing rapid, quantifiable processes to guarantee long term reliability of systems at scale. SRE's contribution here is that where it adds engineering rigor, reliability-oriented metrics, and formalized operational frameworks that align with the principles of software engineering. SRE brings the aspirational goals of DevOps down to concrete, measurable reliability standards that make the overall development and delivery flowing fast, but also reliable, predictable, and shielded against load (Ukis, 2022).

SRE's core mechanism to enhance DevOps is through the strategic use of key reliability metrics such as Service Level Indicators (SLIs), Service Level Objectives (SLOs) and Error Budgets. These are quantifiable thresholds also to monitor and manage system performance in real time. SLIs are critical metrics such as latency, throughput, and availability, while SLOs are targets for this metrics that are acceptable. Complemented with error budget, i.e. threshold for allowed error within given timeframe, SRE allows development teams to make data based decisions of how to prioritize reliability improvements or when to accelerate feature deployment. This structured and feedback driven approach gives an element of risk engineering for teams to keep innovation and system stability in a balance. In addition, SRE also includes practices such as the blameless postmortems, the automated failure detection and the retrospective of incidents which promote a proactive and psychologically safe culture of continuous improvement, system disruptions become opportunities for learning and resilience building.

SRE additionally puts major emphasis on toil reduction and automation, which greatly increases the operational foundations of DevOps. In the realm of SRE, toil is repetitive, manual tasks that don't scale and use up valuable engineering resources. Aggressively identifying and automating such tasks, SRE frees up teams to spend time on innovation and reliability engineering, performance tuning, etc. There are techniques such as infrastructure as code, self-healing systems and real-time observability dashboard that give teams a means to manage these complex distributed systems with as little manual input as possible. What this All this operational evolution does is line up perfectly with DevOps ideals (automation scalability more specifically) only elevating them by making sense of automation in terms of measurement of performance outcome. Thus, the synergy between SRE and DevOps brings maturity, sustainability, and enables an ecosystem for software delivery where agility, reliability, and health of the system come together by design. This integrated model transforms reliability into the engineered property of every release, deployment and system upgrade (Aiyenitaju, 2024).

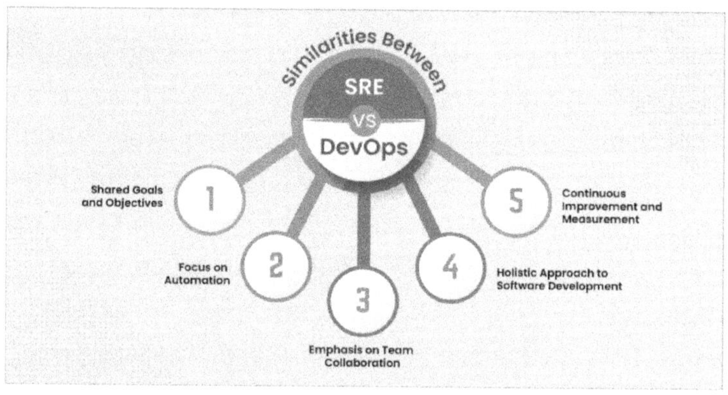

Figure 1.3: Similarities Between SRE and DevOps.

Source: (Veritis, 2022)

1. Shared Goals and Objectives

SRE and DevOps both seek to bridge that gap between the development and operational aspects to deliver a working, reliable, scalable and economical software system. The main objectives they have in common are minimizing downtime, high availability and improving user experience. Both are business driven and help in the rapid, high-quality delivery of software products.

2. Focus on Automation

Automation is a very strong concept both in SRE and DevOps practices. Automation in DevOps is used to accelerate CI/CD pipeline, testing and deployment. It has been used in SRE to reduce toil, manual, repetitive tasks, infrastructure management, incident response and monitoring system. Shared emphasis in this case allows the organizations to scale their operations efficiently with minimal human error and operational overhead.

3. Emphasis on Team Collaboration

The culture of shared responsibility where developers and operations professionals who work together to illustrate mutual goals leads to SRE and DevOps. Through this, it breaks down the traditional silos and encourages sharing of knowledge, faster feedback loop, and an approach that is more integrated for incident management, problem resolution and system optimization. Each discipline values cross functional teams and a blameless culture.

4. Holistic Approach to Software Development

From DevOps and SRE are the views of software delivery and reliability as a continuum of activities from development to test, deployment to maintenance. Both practices bring monitoring, set of performance metrics, and user feedback into the fold of the development lifecycle, integrating monitoring and responsibilities instead of isolating them. The holistic view allows to build the systems that are not only functional but also reliable, secure, and performant in production.

5. Continuous Improvement and Measurement

Oneness with the belief that you have to continuously improve is another common belief between SRE and DevOps. This is done through post incident reviews, error budgeting, service level objectives (SLOs) for SRE and key performance indicators (KPIs), retrospectives, and agile feedback mechanisms for the DevOps. Both are heavily data driven in the sense that they iterate on processes and make their system more reliable and more velocity over time.

Although SRE and DevOps are both from other backgrounds, with different terminologies and implementations, there are many common fundamental principles across these two approaches that are applied in the modern age of software engineering. They both try to bridge the gap between development and operations and to deliver scalable, reliable, high performing systems. Automation is their common focus, improving the efficiency and reducing human error, and the team collaboration strengthens the culture of shared responsibility and knowledge sharing. To become a resilient system, they rely on the holistic view from over the product lifecycle to connect monitoring, user feedback, and performance metrics. In addition, both focus on a continuous improvement using data driven practices such as SLOs, KPIs, post-incident reviews, among others by which they would be powerful allies to attain operational excellence and business agility.

1.3 The Importance of SRE: Ensuring Reliability in Modern Systems

Site Reliability Engineering (SRE) became a foundational discipline in the digital transformation landscape of software systems that serve as the backbone of business operations and serves to ensure a systemic reliability, operational efficiency, and engineering resilience. SRE is rooted in the practices that Google formalized and combines principles of software engineering, systems architecture and infrastructure management to cope with the problems involved in developing and maintaining large distributed, fault tolerant systems (Saurav Bhattacharya, 2024).

At its heart, SRE strives to make an operation system that is scalable and automated, while continuously monitoring and optimizing service reliability, performance, and uptime. Quantitative reliability metrics like Service Level Indicators (SLIs), Service Level Objectives (SLOs) and Service Level Agreements (SLAs) are implemented to achieve this. Such metrics provide the corresponding quantifiable metrics for acceptable system performance and can be used to set error budgets to inform deployment frequency, risk tolerance, and incident response techniques.

Specifically, within modern systems, where the emphasis is placed on the imperatives of cloud-native infrastructure, including container orchestration platforms (such as Kubernetes), microservices ecosystems, and the like, the operational complexity has grown tremendously which naturally presents some rather challenging challenges in itself. Given these environments, Systems are expected to operate at a high availability, low latency, can deal with variable loads and scale as needed, and be able to quickly recover from any partial failures. Thus, SRE adopts automated monitoring, alerting systems, chaos engineering and redundancy mechanisms, as well as predictive analytics to identify the anomalies and prevent the failures in advance – before they can affect the users.

Also, SRE promotes a blameless post mortem culture and continuous improvement, and teams should regard incidents as learning opportunities, not as assigning blame. With observability, automation, incident management protocols, and infrastructure as code (IaC), SRE makes it possible for organizations to operate in a state of operational excellence, and also to facilitate frequent and safe software releases.

Very importantly, SRE functions in DevOps and Agile methodology context as a linchpin between speed of development and up (system stability). With enterprise systems becoming more and more saddled with

Artificial Intelligence (AI), machine learning pipelines, edge computing, real time analytics, the need for SRE only gets more important to ensure that very complex and complex environments deliver consistent, predictable, secure services at scale.

Thus, Site Reliability Engineering is not an operational one but a strategic one in a digitally driven economy today. This ensures that these technological infrastructures are in performant, scalable, and disruption resistant, once again that is adaptable to changing, and can deliver consistent value to users and stakeholders in any industry.

1.3.1 The Impact of Downtime on Businesses

In the era of an increasingly interlinked and real time digital economy, system downtime or any unplanned or planned downtime is threatening to business continuity, performance and profitability. Downtime is the period in which a system, service and application becomes unavailable or fails to operate to the expected levels. Although short interruptions may seem marginal, the cumulative and cascading effects of such interruptions are an operational disruption, financial loss, and long-term reputational damage (Bajgorić et al., 2020).

From a financial standpoint, downtime incurs both direct and indirect costs. These direct costs are loss of transaction revenue, SLA violation penalties, increase in support center volume and resource recovery expenses. Indirect costs are inadvertently reduced productivity, destroyed customer trust, opportunity costs, and long-term damage once again to brand equity. Losses of millions of dollars are possible even in relatively small periods of service unavailability in sectors like e-commerce, finance, healthcare, telecom, logistics etc. Industry research indicates that the average cost of IT downtime is more than $5,600 per minute, which indicates the need for proactive reliability management.

Downtime operationally interrupts the flow of work, data availability and the ability to make decisions. It can stop automatic processes, slow down the customer interaction, break the supply chain, and hinder collaboration among the distributed teams. In the case of the organizations that rely on cloud infrastructure, Internet of Things (IoT) platforms, or AI driven applications, since these depend on real time processing and low latency responses the consequences are further aggravated.

Additionally, from a reputational point of view, frequent or protracted downtimes tend to shrink stakeholder confidence, set this market perception and cause customer attrition. In many such competitive markets, reliability carries a lot of weight as a differentiator. Customer satisfaction and loyalty will noticeably drop down and a further single high-profile outage could bring negative media coverage, regulatory scrutiny and also.

Today, with environments that demand 24/7 service continuity and therefore require no downtime, downtime is a technical failure and a strategic risk. At scale, as organizations rely on more digital infrastructure, minimizing the downtime by means of proactive monitoring, automated failover is a best practice, today but it's becoming a business imperative.

1.3.2 Challenges of Scaling Large Systems

In scaling the large systems, there is a set of multifaceted challenges to be addressed to achieve the operational efficiency, performance consistency and system reliability. With growing size and complexity of software system, the problem of data consistency across distributed components becomes more and more important. In large scale environments, cases like latency, a network partitioning or an inter service communications failure are more likely to occur. In addition, poorly designed and optimized, resource bottlenecks, inefficient load balancing, and state management can cause a degraded system performance. On the contrary, architectures can be made modular, fault tolerant and able to support horizontal and vertical expansion as well as minimizing disruptions to service delivery and they need to be able to scale.

Managing operational complexity at scale is a huge, critical challenge in itself that revolves around being able to observe how things are progressing, which deals with observability. With the growth of systems, more services and dependencies are added, more telemetry data points, thus leading the possibility of detecting anomalies and solving issues quickly gets increasingly more challenging. Large environments are missing in traditional monitoring tools and advanced observability practices such as distributed tracing/real time metrics analysis and intelligent alerting are needed. Moreover, large systems have to be designed for reduced operational toil and automated toil recovery and resilience strategies. To ensure scalability and accommodate the rapid growth of the organization, they need to invest in scalable infrastructure, well designed system and skilled personnel to manage and evolve the complex environment without introducing systemic risk. This shows the need of a disciplined, engineering centric approach to system scaling, especially when it comes to SRE practices.

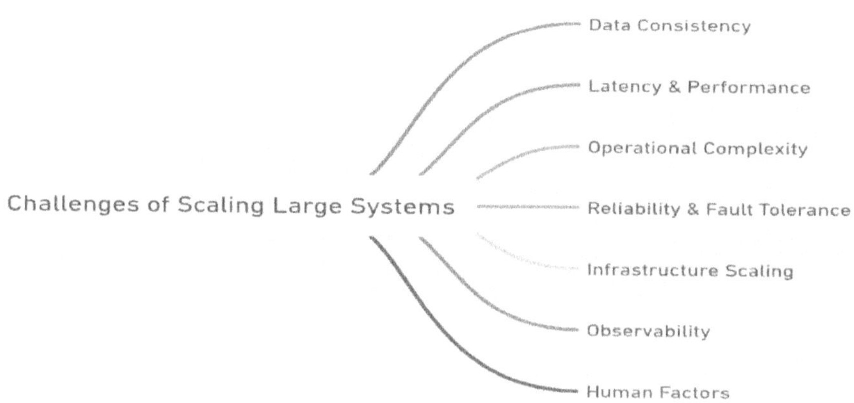

Figure 1.4: Challenges of scaling large systems.

Source: (Self-generated)

1. Maintaining Data Consistency Across Distributed Systems

As systems scale horizontally, data is often distributed across multiple nodes and services. Ensuring consistency in such environments becomes increasingly complex, particularly in the presence of network latency or failure. Large-scale systems must implement sophisticated data replication, consensus algorithms (e.g., Paxos, Raft), and conflict resolution strategies to maintain accuracy and integrity across different system components.

2. Managing Latency and Performance Bottlenecks

With an increase in users, services, and transactions, latency and performance bottlenecks become more pronounced. Asynchronous processing, inefficient queries, and overloaded services can lead to delayed responses or system outages. Identifying and eliminating these bottlenecks requires proactive performance profiling, load testing, and architectural adjustments that support scalability under peak loads.

3. Increasing Operational Complexity

Larger systems introduce greater operational overhead due to the increased number of services, environments, and deployment pipelines. This complexity makes incident detection, root cause analysis, and system maintenance more challenging. To address this, organizations must adopt advanced automation tools, configuration management systems, and scalable observability solutions.

4. Ensuring System Reliability and Fault Tolerance

Scaling a system often exposes hidden weaknesses in its design, particularly in its ability to handle failure scenarios. Fault tolerance must be designed into the system architecture using techniques such as redundancy, graceful degradation, circuit breakers, and self-healing mechanisms. SRE practices play a vital role in validating these strategies through chaos engineering and rigorous reliability testing.

5. Scaling Infrastructure and Resource Management

As systems grow, so does the demand on compute, storage, and networking resources. Efficient resource allocation becomes crucial to avoid overspending and underperformance. Scalable infrastructure, powered by cloud-native technologies and container orchestration platforms like Kubernetes, allows dynamic resource provisioning and better infrastructure utilization.

6. Observability and Monitoring at Scale

Monitoring becomes exponentially harder as the system expands. Logs, metrics, and traces must be collected, correlated, and analyzed in real time to detect anomalies and respond to incidents quickly. At scale, organizations require a robust observability stack that includes distributed tracing, anomaly detection, and intelligent alerting to maintain visibility across the system landscape.

7. Human and Organizational Constraints

Large-scale systems also present challenges related to human cognitive load and cross-team collaboration. The increased scope of services makes it difficult for individuals or small teams to maintain full situational awareness. Clear ownership models, documentation standards, and cross-functional communication channels are essential to avoid operational silos and inefficiencies.

Solving for large systems is both a technically and a business problem, with kernel arguments about how to handle apps and data on large systems being argued by multiple actors at once. Reliability of the system is ensured by ensuring fault tolerance, optimizing infrastructure resources and maintaining observability at scale. At the same time human and organizational constraints need effective collaboration, clear ownership and automation. And for the practice of contemporary SRE these areas are key to address as they are critical for sustainable growth.

1.3.3 The Cost of Unreliable Software

In an increasingly interconnected and digitized world the price of unreliable software is no longer a lost opportunity when the software can't release the results but extends now to critical operational, reputational, and financial measures. The impact of such issues grows larger and larger as more and more of those services are delivered over software driven infrastructures. Software predictability on the other hand ensures business continuity, integrity of service and a significant measure of stakeholder confidence in whatever venue; i.e., in consumer applications, enterprise systems or embedded software that reside in devices. One of the industries that are particularly vulnerable are banking, healthcare, e-commerce and telecommunications where every transaction and interaction have great value. For such sectors, a short period of downtime would mean a loss of revenue, a missed business opportunity, a regulatory noncompliance and would drop customer satisfaction (Evans, David.S., Hagiu Andrei, 2019).

Unreliable software has financial repercussion that are substantial and complex. The direct losses is typically slippage of SLA (Service Level Agreement), transaction failure, remediation cost, and customer compensation payout. There are, however, indirect and no less harmful consequences which include damage to brand equity, lost customer trust, and negative media coverage. The decline in market valuation for companies with frequent or high profile outages is a common consequence of reducing the level of investor confidence. In many cases these incidents quickly become reputational crises from which it will take years to recover. This is especially true in today's social media centered society, and where negative experiences will spread quickly, even one high impact failure can take a turn for the worse when it comes to customers as well as the perceived public trust.

Another hidden, but critical source of cost is inefficiency resulting from the unreliability. Firefighting with unstable software is a diversion of engineering teams from their strategic initiatives and innovation. Incident response, root cause analysis, patching and manual intervention tasks, often known as toil in SRE practices, are burdensome to them. Not only does this incur additional operational costs, time to market is increased, product development cycles are slowed down, and a growing technical debt backlog build. The demand of keeping those unreliable systems up has given staff engineers nothing but burn out, low morale and high attrition rates. Such workforce challenges make the unreliability problem even worse, complicating the feedback loop that results in organizational in agility and performance in the long term.

As far as user experience goes, modern consumers have little patience for software that doesn't deliver, and there is an increasing demand for such software. In the digital age, reliability is now a core differentiator, as with digital experiences, you form a large part of customer engagement. Applications must be responsive, performant and available 24 x 7. This norm can be deviated from either in the form of latency, intermittent failures or full outages and users can choose to abandon the service. This means reduced customer retention and lifetime value for a consumer facing application. This means that for enterprise solutions, clients will be dissatisfied and the contracts terminated. The implications are worse in the regulated industries due to the unreliable software. Breach of data privacy laws, regulatory violations, mandatory audits and huge fines are possible. In worst case, they very real legal risks, especially when, for example, they compromise customer data or impact safety critical systems.

Site Reliability Engineering (SRE) directly tackles these challenges by putting reliability on the first-class list of concerns in software engineering and operations. Instead of writing reliability as an afterthought, SRE incorporates reliability into the software soak cycle via mechanisms like service level objectives (SLOs), error budget, and observability. However, these tools allow teams to create and characterize acceptable thresholds of system behavior and check how the components in the system behave toward them. SRE also teaches diligent incident management and continuous improvement methodology such as blameless postmortem and continuous improvement loops to lesson out failures to learning pain rather than learning liability. Also, it pushes the idea of reducing the human toil via automation and engineering solutions for scalability and fault tolerance.

In the end, SRE brings system reliability from a metaphorical pursuit to a quantifiable and enforceable business metric. It enables organizations to provide consistent and high-quality digital services at efficiency and user trust. For all the software is the backbone of value delivery in a competitive business environment, it is not just a technical attribute; it is a strategic imperative. Adopting SRE principles allows companies to pay less for unreliable software, increase the robustness of systems, and be resilient and grow in the digital age.

1.4 Key Principles and Objectives of SRE

A modern operational mindset for engineering site reliability, SRE uses a set of software engineering approaches to build and run scalable, highly reliable systems. SRE is a philosophy that operations is a software problem. SRE looks toward employing engineering discipline on the operational tasks and enabling systems to scale reliably without excessive reliance on reactive measures and manual processes (Betsy Beyer, Niall Richard Murphy, David K. Rensin, Kent Kawahara, 2018).

SRE is based on measurable and actionable metrics, proactive approach towards incident response and extreme passion towards continuous improvement. SRE model aims at facilitating the bridging between development and operations by putting their goals on the same page: the fast deployment of features and high degree of system reliability. They are not merely a methodology but also a culture shift that allows organizations to operate resilient services at a high speed of development.

Key Objectives of SRE:

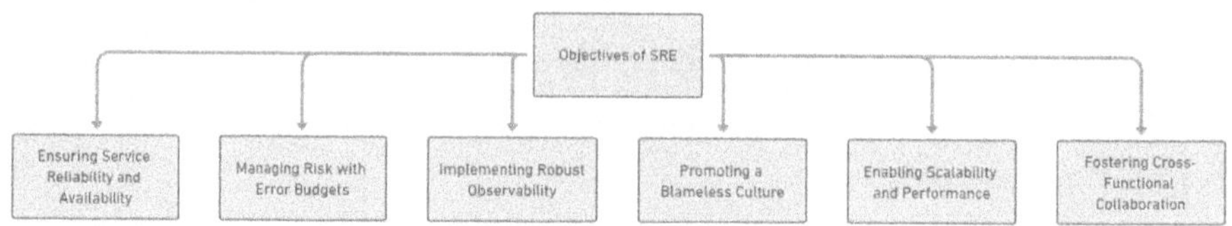

Figure 1.5: Objectives of SRE.

Source: (Self-generated)

1. **Ensuring Service Reliability and Availability:** Site Reliability Engineering (SRE) is one of the main things that are trying to keep those two objectives in alignment, service uptime and performance, with the business requirements and user expectations. It includes setting up and observing well defined Service Level Objectives (SLOs), constant measurement of error rates and system downtime reduction through proactive actions. To achieve such strategies, redundancy is implemented, graceful degradation techniques are utilized, and self-healing mechanisms are integrated into the system architecture.
2. **Automating Operational Tasks (Reducing Toil):** These repetitious, manual and low value tasks or the "toil" consume valuable engineering time. The goal of SRE is to discover such toil and do away with it through automation. Deploys, scaling, incident response and system recovery are all automated to increase efficiency and decrease the human error.
3. **Managing Risk with Error Budgets:** Error budgets are introduced as a way to trade off between innovation velocity and reliability in SRE. The error budget is the amount of system unreliability that is allowed within a given time period, and represents a boundary within which development teams can experiment and release new features in knowing their systems won't be destabilized.

4. **Implementing Robust Observability:** There is a need to observe the state of the system. It is SRE's responsibility to make sure that systems are instrumented to collect logs, metrics, and traces for real time understanding of the system's health and performance. This allows SRE's to quickly figure out anomalies, do root cause analysis and resolve incidents better.
5. **Promoting a Blameless Culture:** An important cultural principle of SRE is to treat failures as opportunities to learn. Teams that conduct blameless postmortems find flaws in the system as well as the process without blaming the people. It helps create a psychologically safe environment of continuous improvement.
6. **Enabling Scalability and Performance:** In a system, to grow is to need to scale. The main focus for SRE is to design and maintain the architectures to be able to sustainably service increasing loads as the business demands. Capacity planning, performance tuning and stress testing are invoked to make sure systems work in par under various scenarios.
7. **Fostering Cross-Functional Collaboration:** This brings SRE to encourage collaboration between development and operations teams. In shared responsibilities, clear communication and joint ownership of reliability metrics, technical efforts and business goals are aligned more quickly and an engineering environment is more cohesive.

Sites Reliability Engineering (SRE) core principles and objectives are driven by making systems as reliable as possible, automating monotonous tasks, spending error budgets, and making observability and collaboration everyone's responsibility. SRE uses software engineering practices to integrate them into operations and make services available, decrease toil, and scale systems efficiently. It also encourages a blameless culture in which the focus is on learning from incidents through continuous improvement. In the end, SRE allows an organization to balance new ideas as much as possible with the need for stability, turning reliability into a measurable and strategic business asset.

1.4.1 Service Reliability as a Core Focus

Service reliability is not one of many performance metrics in Site Reliability Engineering (SRE); it is the central principle around which all system activities are built. This is based on the fact that modern software systems are critical missions, and any small disruption will incur heavy losses, damage to reputation, and diminished user trust. Because of this, SRE redefines reliability from a reactive concern that we only think about when the system is down to a proactive design goal that influences every decision in the life cycle of the software from development to operations (Adolph, 2016).

A holistic and forward-thinking approach is required when aiming for this level of reliability. SRE practitioners are expecting failure, not stability, and thus design systems. As a consequence, redundancy, failover mechanisms and self-healing architectures are deliberately incorporated in order to ensure that services can still operate even during stress or under component failures. Graceful degradation strategies are also implemented to keep users with partial functionality in case of adverse conditions as opposed to total service outages.

One of these is proactive failure detection and rapid incident response. Instead of waiting for users to complain about problems, SRE teams create systems to detect anomalies in real time, on give visual and auditory warning, and trigger automated or semi-automated process to terminate the faulty state. To discover system weaknesses and verify the fault tolerance mechanisms, chaos engineering practices like deliberately introducing faults in the controlled environments are used. Fairly, practices of these kinds not only enhance resilience but do encourage a learning culture and continual improvement.

The reduction in manual operational toil, or repetitive, low value tasks, is also critical to ensure reliability as they distract from higher order engineering work. These tasks are automated using tools and scripts according to SRE, and it removes the human error by reducing the number of team members working to focus on reliability engineering and strategic system improvement. System health checks, log analysis, verification of deployments and scaling of resources is done using automation and this sustains high availability and consistent performance.

Apart from technical safety, SRE promotes a cultural shift from the blame culture to blameless postmortems and shared responsibility. If failures do happen, teams are not blaming people but instead are trying to discover latent system weaknesses. It creates environmental for transparency, psychological safety and collaboration which just enables teams to tackle the root causes effectively and to learn from incidents in constructive way. Then the insights are used to strengthen the system architecture, revise the operational runbooks and updating the disaster recovery procedures.

SRE positions reliability as a core engineering value, and therefore a business enabler, by putting it at the core of both engineering and organizational values. It makes systems available, and in good (but not optimal) performance, but also predictable, scalable, and maintainable. Due to the times we live in today, being down means being out, and losing revenue is directly attributed to poor user experience and position relative to competition in the fast paced digital economy.

In the end, by focusing on service reliability as the key, organizations can innovate regardless of scale and complexity, operate at the highest levels, and provide continuous value to users.

1.4.2 Measuring and Managing Risk in Systems

In modern distributed systems, risk is an inherent element of both design and operation. Site Reliability Engineering (SRE) approaches risk not as something to be entirely eliminated, but as a quantifiable and manageable dimension of system behavior. Understanding and addressing risk allows organizations to maintain a careful balance between system reliability, innovation, and business agility.

Effective risk management begins with the identification and classification of risks. These may include technical risks such as system downtime, latency spikes, data inconsistency, or deployment failures; operational risks such as human error, insufficient monitoring, or capacity shortages; and business risks such as breach of service-level commitments or regulatory non-compliance. SRE emphasizes the use of structured methodologies to identify potential failure modes and assess their likelihood and impact on system availability and user experience.

One of the key ways SRE teams manage risk is through risk acceptance frameworks that align technical tolerances with business goals. This includes determining how much unavailability or performance degradation is acceptable under specific conditions. Instead of striving for theoretical perfection, SRE encourages teams to define clear reliability thresholds and prioritize efforts based on potential business impact. This pragmatic approach ensures resources are focused on the most critical areas, avoiding unnecessary over-engineering and optimizing cost-effectiveness.

Risk management also involves prioritizing and mitigating high-impact risks using engineering strategies such as redundancy, fault isolation, failover mechanisms, and automated rollback procedures. Resilience is built into the system so that when failures do occur, their scope is limited and recovery is swift. Techniques like chaos testing and load simulations are employed to stress the system under controlled conditions, allowing teams to observe how it behaves and make improvements before issues arise in production environments.

Additionally, post-incident analysis and learning play a vital role in risk management. Every incident is treated as an opportunity to uncover latent system weaknesses and refine existing defenses. By conducting blameless postmortems, SRE teams create a culture of transparency and continuous improvement, leading to reduced risk exposure over time.

Another key aspect of measuring risk involves maintaining visibility through observability tools. Real-time monitoring, logging, and tracing systems provide actionable insights into system health, enabling teams to detect anomalies before they escalate into full-blown incidents. These insights support proactive risk mitigation, capacity planning, and informed decision-making.

Ultimately, measuring and managing risk in systems is about enabling systems to evolve and scale safely while maintaining user trust and service continuity. Through structured risk evaluation, thoughtful design, and proactive operational strategies, SRE helps organizations deliver reliable services in an inherently uncertain technological landscape.

1.4.3 Automation Over Manual Intervention

'Automating everything over manual operations', is probably one of the foundational principles of Site Reliability Engineering (SRE). Both to scale up size and increase complexity, it is no longer efficient, error prone, or sustainable to rely on human intervention. Automation gives engineers more time to devote to the system design, optimization and innovation aspects of their work (Ricky Smith, 2011).

With an SRE framework, repetitive, repetitive, and time-consuming activities are often identified as "toil" which is systematically targeted for automation. Some of these include configuration management, deployment processes, infrastructure provisioning, monitoring and incident response. Since these functions can be automated, organizations can therefore save on operational overhead, eliminate variability in outcomes and achieve swift and correct execution of critical procedures.

On the other hand, critical automation is also present in incident mitigation and recovery. Systems are automated and have self-healing mechanisms such as auto scaling, service restarts, and circuit breakers. They can recover from failures with little or no human involvement. Such a capability for rapid response has a dramatic effect on system availability and user experience.

Moreover, the CI/CD pipelines are a critical area where automation brings great value. Automated build, test and deployment workflows allow code to be safely released extremely fast with no compromise in veracity. Besides, this also shortens time to market and reduce bugs in production.

Also, IaC allows teams to manage and provision computing environments from machine readable configuration files. This way enables us to achieve repeatability, version control, peer review, and the deliberate, auditable, and, if needed, reversible changes of infrastructure.

The turn from reactive system maintenance to proactive is enabled by making automation sit at the heart of operational strategy: SRE. This offers reduction of human error and an addition of predictability giving us the ability to run systems at scale reliably. In the end, automation is not simply a technical preference but an essential strategic factor for maintaining high reliability, scalability and operational excellence in the digital realities of the modern digital world.

1.5 The Role of an SRE in an Organization

Modern organizations desperately love SREs precisely for their involvement on overlapping and largely bifurcated topics: software development and IT operations. Their main task is to make sure those complex software systems are scalable, reliable, and performant enough, and this used to be true, especially when organizations move towards cloud native and microservices-oriented architectures. In traditional systems, System Administrators (Sysadmins) were responsible to maintain systems and deal with incidents, whereas SREs work in a proactive, engineering-forward way, focussing on automation, monitoring and optimization for performance with the purpose of minimizing human intervention, and maximizing uptime (Wilensky, 2015).

SREs are not operators, they are builders. Tools, code and systems are architected for repetitive tasks to be eliminated, failure impact reduced and service delivery accelerated. Their work ensures that even with extreme load, these services can meet performance benchmark, can still be observable, can quickly recover from failures and can scale dynamically without impacting user experience or business continuity at any point in time.

Figure 1.6: The role of an SRE in an organization.

Source: (Self-generated)

1. Bridging Development and Operations

SREs bridge the cultural and technical divide between the development and operations teams through practice of deploying smoothly, getting faster feedback loops, and building greater service stability. By aligning it with the production readiness criteria, this makes reliability a shared responsibility and not something that is an afterthought.

In most cases, SREs have to do design reviews and deployment planning with development teams to project potential risks and bottle necks. As a result, it is about more resilient applications and services, that are designed with reliability and scalability at the core of their architecture.

2. Incident Management and Response

Incidents and service availability are one of the most visible SRE roles. SREs are the first responders for outages, or poor performance degradations. With incident response protocols that are structured, playbooks being automated, and observability tools, they are able to quickly identify root causes and bring the service function back.

SREs go beyond firefighting: institutionalizing the learnings from these incidents through blameless postmortems to make sure that the knowledge learned is shared and not used for blame." It is this practice that turns system failures into chances for systemic improvement.

3. Performance Optimization and Capacity Planning

Performance profiling, stress testing, load forecasting is SREs responsibility. As they constantly monitor system behavior and understand how users are using the system, they can decide on the strategy of infrastructure provisioning and scaling.

Just like any other planning, capacity planning is a key function of SREs to anticipate the future demand and design the system to take load spikes. It prevents costly downtimes and keeps the system at its peak performance as traffic grows rapidly or in season.

4. Automating Operations and Reducing Toil

The expression toil acts as a cornerstone of SRE philosophy and refers to any manual, repetitive and automatable operational tasks. Wasting time alone is one of the things that toil achieves: it wastes time, and the more time that is wasted, the greater the likelihood that a human error will occur. They are responsible for the automation pipeline for deployment, monitoring, testing and recovery.

Amongst some of the tools and techniques that SREs deploy to streamline operations are infrastructure as code (IaC), CI/CD pipelines and automated rollback mechanisms. This automation means not only for increased operational efficiency, but also for increased time for innovation and engineering excellence.

5. Building and Maintaining Observability

SREs also make sure the systems are observable, so they are observable at any time, that they are observable in the sense that they are monitorable, diagnosable, and understood. The word observability goes a bit further than just monitoring, and includes solutions for setup for a comprehensive logging, distributed tracing, real time dashboards, anomaly detection and intelligent alerting systems among many other things.

An observable system translates informative data into actionable insights, which the teams use to prevent the issues from worsening. So SREs work very closely with product and infrastructure teams to introduce observability into the workflow and keep a feedback loop between system performance and engineering action.

6. Enforcing Reliability Engineering Standards

Defining reliability objectives, setting error budgets and getting reliability testing into CI/CD pipelines are all things that SREs help foster through a reliability first mindset across the organization. We are not including SLIs and SLOs in the context of this, but other mechanisms include:

- Resilience Testing
- Game Days (failure injection testing)
- Uptime and Latency Threshold Audits

Quantification and enforcement of system expectations are used. Thus, SREs make the reliability measurable and actionable, the goals of engineering and business always align.

7. Cross-Functional Collaboration and Cultural Impact

SREs are the enablers of the cultural change within the organization. They encourage the spirit of a shared accountability, learning and thinking in systems. Over development and QA, product management, and business operations, they work as one.

This cross functional collaboration is essential for the breaking down the silos and enhancing transparency, to allow the variations in the work, to be seen by the entire stakeholder and to also understand how his work helps build a dependable system. Typically, SREs provide on call support, mentor teams, help elicit internal best practices and share that in runbooks, playbooks and documentation.

2. **Ensuring Compliance and Governance**

SREs exist in industries that are regulated by standards (finance, healthcare and telecommunications) and guarantee that the systems are designed and operated according to internal policies and external standards. That includes audit trails, uptime SLAs, and in fact, access control mechanisms specifically.

Here SREs work with the compliance teams to create a governance framework that balances security, traceability, and accountability without bringing in any extra friction to the development and operation process.

1.5.1 SRE Team Structures in Different Companies

Site Reliability Engineering (SRE) has evolved into a critical discipline across organizations of varying sizes and industries, but the way SRE teams are structured often depends on the company's scale, maturity, technical infrastructure, and cultural readiness. While the foundational principles of SRE remain consistent—such as ensuring reliability, automating operations, and managing system performance—the organizational implementation of these principles can vary greatly. Broadly, companies adopt different models of SRE team structures, including centralized teams, embedded teams, hybrid models, and platform-oriented SRE groups.

In startups and small-to-medium enterprises (SMEs), SRE teams are often lean and operate in a centralized model. A small group of reliability engineers supports multiple product and infrastructure teams, focusing on standardizing observability, incident response, and automation. Given resource constraints, these SREs typically wear multiple hats—acting as both infrastructure experts and reliability consultants. Their primary focus lies in setting up foundational systems, ensuring monitoring coverage, building deployment automation, and putting incident response protocols in place. These centralized SRE teams often serve as "enablement" teams, helping developers own their services more effectively while driving reliability practices organization-wide.

On the other hand, larger technology companies like Google, Netflix, and Amazon often utilize an embedded or hybrid SRE model. In this structure, SREs are assigned to specific product or service teams where they work closely with developers to design, build, and maintain reliable systems. This close proximity allows for deeper understanding of the application architecture, faster feedback loops, and shared ownership of service reliability. These embedded SREs act as domain experts, introducing practices such as failure mode analysis, game days, service-level management, and infrastructure optimization tailored to their specific teams. In some organizations, hybrid models are adopted, where a core SRE platform team supports embedded SREs by building common tooling, standardizing practices, and ensuring alignment with broader reliability goals.

When a company is digital transforming or operating in a heavily regulated environment (finance, healthcare etc.), SRE teams that are created to balance risk management, compliance and audit readiness. It may split these teams into subteams capable of things like observability, security, performance testing, capacity planning, or compliance automation. In such environments, SREs don't just keep an eye on the service health and uptime, they also need to follow strict service level agreements (SLAs), data protection standards and failover strategies. Their team structure may integrate with DevSecOps and compliance teams to ensure that the reliability efforts are coordinated with legal and operational requirements.

There is also the Platform SRE model, which is where the team is charged with keeping the shared infrastructure platforms like Kubernetes clusters, CI/CD pipelines, and logging systems reliable. These SREs create reusable solutions as well as self service platforms for product teams, allowing reliability without SRE's being involved in each service. As the number of microservices in an organization is high, this model is particularly useful because embedding SREs in every team would not be scalable. The SREs on Platform are tasked with optimizing developer's workflows, increasing automation and enforcing global reliability standards.

But the most important thing to understand is that the structure of an SRE team is rarely static and evolves as the organization matures. For a company, a first model can be a central one, and then they adopt a hybrid or platform model as sense of systems and services complexity increases. As SREs become more mature, it walks a line between the other operators and developers, particularly when adoptive philosophies like 'you build it, you run it' begin to be supported by strong internal SRE consulting functions.

In the end, there is no one size fits all way to structure SRE team. Instead, the SRE structure for the successful organizations aligns with their technical architecture, with team autonomy, with operational maturity, with business goals. It doesn't matter what the structure of your organization is, the goal is the same: improve reliability of the system, reduce toil in operations, and give teams the means to deliver high quality software at scale.

1.5.2 Skills Required to Succeed in SRE

In order to be a successful Site Reliability Engineering (SRE) professional, professionals need to have a variety of technical, operational, and interpersonal skills. SRE is a multidisciplinary field that combines notion of software engineering, systems administration, and IT operations and it requires the practitioner to be equally capable of coding, infrastructure management and reliability engineering. To understand how complex, distributed systems behave and be able to evaluate performance, reliability issues, one must have a solid foundation in the computer science fundamentals, namely data structures, algorithms, operating systems, and computer networks.

For an SRE, programming and scripting proficiency is one of the most crucial skills. Automating operational tasks, building internal tools, managing infrastructure are done using languages such as Python, Go, Java, Shell. SRE is about the backbone of automation, they need to be able to minimize the amount of manual toil with the help of writing scripts and building self healing systems. Configuration management and infrastructure as code practices of using tools like Terraform and Ansible as well as Chef allow for scalable, repeatable and consistent deployments and management of the infrastructure.

The table below brings down the diverse skills required in SRE and the key skill areas, and how they are important.

Table 1.2: Core Skill Sets for Successful Site Reliability Engineers (SREs).

Skill Category	Examples	Purpose
Programming & Scripting	Python, Go, Shell, Bash	Automate tasks, develop internal tools, reduce manual toil
Infrastructure Management	Terraform, Kubernetes, Docker, Ansible	Manage scalable infrastructure, container orchestration, infrastructure as code
Monitoring & Observability	Prometheus, Grafana, ELK Stack, Open Telemetry	Enable system visibility, anomaly detection, and proactive incident handling
Cloud Platforms	AWS, Google Cloud, Azure	Build, scale, and manage cloud-based distributed systems

Skill Category	Examples	Purpose
Security & Compliance	Encryption, IAM, vulnerability scanning	Protect systems and ensure regulatory compliance
Soft Skills	Communication, collaboration, incident response, critical thinking	Coordinate across teams, manage outages, conduct blameless postmortems
Capacity & Performance Planning	Load testing, stress testing, auto-scaling, traffic forecasting	Maintain performance and prevent bottlenecks during high loads

Source: (Self-generated)

Another core competency of an SRE is system observability and monitoring. It requires engineers to be able to design and maintain such monitoring systems using tools such as Prometheus, Grafana, and the ELK stack. This observation ensures availability of logs, metrics and traces for real analysis and thus helps in faster issue identification, root cause analysis and incident resolution. This capability allows reduced MTTD and MTTT, which are critical for high availability systems.

A competence in cloud native infrastructure is equally important. SREs are required to know cloud environments and services, such as compute instances, load balancers, virtual networks, storage solutions, etc. In addition to skills in Kubernetes, containerization, CI/CD pipelines, system scalability and maintainability are further enhanced. With hybrid or multi cloud strategies often adopted by the organizations, SREs must have the skill of working across diversified infrastructure landscape.

From a security standpoint, SREs must make sure that they integrate secure practices into system design and the daily operations. As such, it includes access control, secure data transmission, regular vulnerability assessment and a compliance with data protection laws. However, particularly in regulated industries, it is important to understand how to achieve these standards such as, SOC 2, ISO 27001 or HIPAA – so that your operations will be reliable and secure.

In addition to the technical, the strong soft skills are indispensable. SRE often work with the development teams, the product managers, the QA engineers and the business stakholders. But it also has to be effective, and be effective communications whether that is to explain outages in a complex system or to write clear documentation or to do blameless postmortems. Managing high stakes incidents requires stress management, decision making under high pressure, teamwork and so on.

In addition, a successful SRE can also plan for the growth of the system. Capacity planning is the process of forecasting the need, evaluating the current system threshold and supplying the resources accordingly. It prevents overloads in the system and is prepared for the maximum amount of traffic. SREs bring strategic contribution to scalability and efficiency through their support to performance profiling, stress testing, and resource forecasting.

Finally, an SRE needs to be a continuous learner and improvement. Technology space keeps changing with new tools, techniques and practices ready to greet it all the time. The SREs must be updated about automation, reliability engineering, DevOps, and artificial intelligence in operations (AIOps) developments to be ahead of the challenges coming their way.

To sum up, being an SRE requires a person with a good technical depth, operational knowledge, and strong collaboration skills. The resultant skills of these combined skills allow organizations to construct highly resilient systems that provide always high levels of performance, availability and user satisfaction.

Multiple Choice Questions (MCQs)

1. **What was the primary organization behind the development of Site Reliability Engineering (SRE)?**

   A) IBM
   B) Microsoft
   C) Google
   D) Amazon

2. **Which of the following best describes the core philosophy of SRE?**

   A) Increasing complexity to test system limits
   B) Automating repetitive tasks to improve system reliability
   C) Eliminating the need for developers in operations
   D) Reducing costs through outsourcing operations

3. **How does SRE differ from traditional IT operations?**

   A) Avoids interaction with developers
   B) Prioritizes automation and service-level reliability
   C) Emphasizes manual ticket-based workflows
   D) Focuses solely on hardware provisioning

4. **What is one key enhancement SRE brings to DevOps?**

   A) Introduces service-level objectives (SLOs) and error budgets
   B) Focuses only on coding practices
   C) Shifts testing to end-users
   D) Eliminates the need for monitoring

5. **Why is SRE critical in modern systems?**

   A) It improves physical security of data centers
   A) It eliminates the need for capacity planning
   B) It reduces marketing costs
   C) It ensures high system reliability and minimizes downtime

6. What is a major challenge in scaling large systems?

A) Shorter release cycles
B) Maintaining data consistency across distributed systems
C) Simplifying software architecture
D) Reducing internet speed

7. Which of the following is a financial impact of unreliable software?

A) Loss of stock market confidence
B) Better customer loyalty
C) Increased hiring
D) More code reuse

8. What principle does SRE promote to manage risk effectively?

A) Avoiding change
B) Frequent unplanned outages
C) Strict manual oversight
D) Proactive failure testing and risk assessment

9. Why does SRE emphasize automation over manual intervention?

A) To reduce code quality
B) To eliminate human error and scale operations
C) To improve user interface design
D) To reduce the need for infrastructure

10. In an organization, what is one responsibility of an SRE team?

A) Writing press releases
B) Hiring software testers
C) Creating branding strategies
D) Performance profiling and capacity planning

Answers

1.	2.	3.	4.	5.	6.	7.	8.	9.	10.
C	B	B	A	D	B	A	D	B	D

CHAPTER 02

Building Reliable and Scalable Systems

2.1 Designing for High Availability and Fault Tolerance

It is expected that the modern digital systems offer seamless and continuous service no matter the unpredictable challenge like hardware malfunctions, power outages, or cyberattacks. The recent development of digital infrastructure is becoming an integral part of most industries, including e-commerce, banking and healthcare, as well as cloud computing, hence the importance of system resilience. In today's world user expect 24/7 availability, always on and always available, as well as always performant performance, without any impact of external disruptions or attacks in the devices or networks connecting to the Cloud. This means that the organizations have to move from reactive to proactive architectural planning for constant uptime (Religions, 2009).

A system design approach to high availability (HA) is to eliminate the single points of failure and bring the downtime to the barest minimum. It is the deployment of redundant hardware components, geographically separated resources and fail over mechanisms which are activated when a component or service becomes unavailable. It is not just about duplications but about first thinking how the resources will be orchestrated, monitoring them automatically and switching between primary and secondary systems without a hitch. Further enhancement of HA in cloud environments can come by running on scalable and elastic infrastructure which scales up and down automatically depending upon the workload demands.

However, fault tolerance is concerned with constructing systems which operate correctly even in the presence of failing components. It means more than giving hardware redundancy, but more software strategies including the exception handling, service isolation, and recovery logic. Instead, applications must be designed to degrade gracefully, in which they can lose some non critical functionalities, but that they still provide core services. For instance, a video streaming platform may disable personalized recommendations (but keep serving video streaming) until a backend service comes back upon failure, instead of letting users keep watching the video, but without personalized recommendations.

Oxygenation of these principles also has to be done by the organizations and this includes the adoption of intelligent infrastructure management practices such as automated load balancing, health checks, and Chaos testing. The load balancers guarantee even distribution of user traffic across resources and rerouting the traffic back from the unresponsive nodes. These health monitoring tools gives you real time visibility into system performance while chaos engineering tests that how the system is robust. Together, these strategies build a community where systems are designed resiliently and as such, continuous service delivery is ensured and the impact of extraordinary events is kept to a minimum.

2.1.1 Understanding High Availability Architecture

High Availability (HA) is a system's ability to stay operational for a desirably long period of time in the realm of distributed systems and enterprise IT infrastructure. In other words, HA is not just a byding, HA is the consequence of making particular architectural choices, which, in their turn, imply service continuity, minimal downtime and effective fault management. HA can be expressed from a mathematical point of view in terms of "nines" availability, for example, 99.9% ("three nines") or 99.999% ("five nines") uptime and that implies that HA hours are only a few minutes per year. It works out as a business critical metric to many sectors including finance, healthcare, and cloud services, because downtime can result in loss of data, stoppage of operations, or revenue loss (Makarova et al., 2020).

System redundancy is the key to high availability, in which the key components and subsystems of a system are duplicated so that failure of one does not bring the system down. Computes nodes, network paths, storage arrays, power supplies, and cooling systems are included in these. Most HA strategies are based on architectures like active-active (all node process traffic concurrently) and active-passive (backup system is on standby until failure is detected). Failover mechanisms that ensure that these topologies do not suffer from system anomalies by monitoring the system health with heartbeat messages and seamlessly transition on the anomalies to save the users sessions and service state are underpinned.

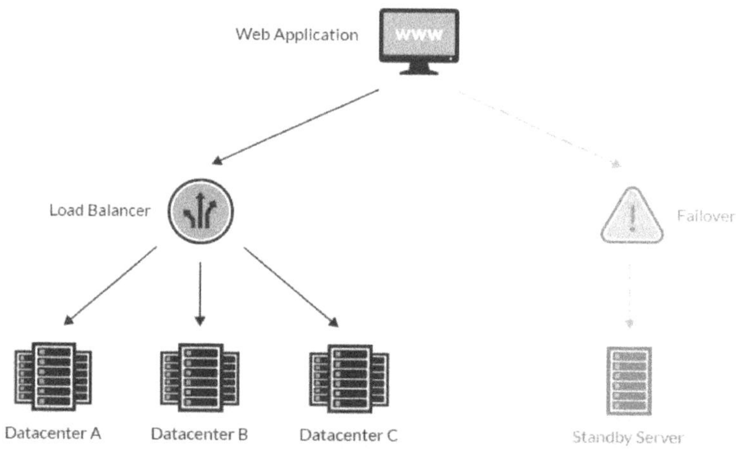

Figure 2.1: High Availability Architecture with Load Balancing and Failover.

Source: (Imperva, 2021)

From a theory of distributed systems point of view, HA also depends on data replication strategies and consistency models. For example, data in distributed databases and replicated file systems must be replicated across multiple nodes based on the two possibilities of synchronous replication (strong consistency), or asynchronous replication (eventual consistency). Consistency and availability in such a unreliable network is maintained through consensus algorithms like Paxos, RAFT and Zab. This helps with leader election and resolution of conflicts in clustered environments because these algorithms make sure that a majority (quorum) of replicas agree before committing a system state.

Additionally, geographical fault isolation is a huge factor in high availability by dispersing system resources across multiple Availability Zones (AZs) and Regions. It guards against localized failure such as natural disasters, regional power outage, as well as failure of datacenter specific hardware. This principle is baked into all of the cloud service providers such as AWS, Azure, or Google Cloud where the infrastructure is designed with HA as a default and HA is achieved through cross AZ or cross region replication, automated failover and defined RTO and RPO based disaster recovery configurations.

The other major pillar of HA architecture is observability and real time telemetry. System health is inferred though continuous monitoring of metrics (CPU usage, memory utilization, disk I/O), logs (event and error logs), and traces (request paths), and anomaly is detected in advance. Modern monitoring platform usually contains an AI/ML based anomaly detection to detect faulty prior to leading to failure and proactive intervention. Prometheus, Grafana, New Relic, and other tools provide visibility into system behavior and allow implementing self healing systems where failed containers or services are replaced or restarted with no human intervention by means of automated scripts or orchestration layers such as Kubernetes.

Chaos engineering is finally becoming the scientific method of empirically testing and improving HA architectures. Building upon this idea, engineers can devise their own injections by killing services, adding latency, corrupting data to see how the system responds and identify failure isolation and recovery logic points which may be weak. The system resilience under stress conditions is validated and the failover policies, retry strategies and rollback mechanisms are refined using this approach. Together, these scientific principles—redundancy, consensus, replication, observability, geographic distribution, and failure injection—form the backbone of high availability architecture in modern computing environments.

2.1.2 Building Fault-Tolerant Applications

Fault tolerant applications are meant to work correctly despite of hardware malfunction, software faults, network outages or human errors. This resilience is built through design decisions that are anticipatory of failure as an expected behavior as opposed to an incident. The hallmark of a fault tolerant system architecture is that when a component defects on graceful degradation, that is, the rest of the components in the system are not required to fail abruptly or completely, but instead degrade completely or reduce service to the user to work partially or in a degraded mode at reduced capacity or in reduced capability. It becomes critical for us for applications requiring very high reliability and availability, which include many applications where high reliability is needed, such as in aviation control and medical monitoring, as well as processes such as financial transaction processing (Israel Koren, 2020).

Redundancy at the application level is one of the main techniques in building fault tolerant systems. This also involves duplicating services and processes in case one fails, and the other takes over seamlessly. For example, in microservices architecture, redundancy is present by running multiple instances of a service behind a load balancer and failover happens automatically. In large system, communication between microservices can be managed by the service meshes such as Istio or Linked, detect the service failures, and redirect the traffic without involvement of the developers. The architecture of this approach guarantees system uptime despite disruptions to some components.

MCM also provide error detection and correction mechanisms that were essential components of fault tolerant applications. Checks should exist within the software to identify anomalous behavior through built in checksums, validation logic and heart beat signals between distributed components. Consequently, such recovery strategies as retry logic, circuit breakers, and fallbacks are generally used to achieve it. One example of this is if a service is temporarily unavailable a circuit breaker could prevent sending any further requests to the failing component so that the rest of the application can continue working without cascading failures. At the same time, fallback procedures such as serving cached data are put in place to keep the user going, in case things go wrong.

Stateful applications have their own requirements when it comes to fault tolerance. Due to the central role of databases and message queues in most systems, there is a need to replicate data across nodes so that data is consistent and does not get lost. WAL and snapshotting ensure that committed data can be replayed or restored after crashes using techniques such as WAL and snapshotting. Another fault tolerant pattern is event sourcing where each change in state is recorded as events that can be replayed to restore a system's state from event logs in the case of catastrophic failure.

In addition, application fault tolerance also includes separation of components by implementing asynchronous communication. Kafka, RabbitMQ, Amazon SQS and more serve as buffers between services serving to keep a part of the system running even if the one down. By doing so, not only resilience is increased but better scalability and performance under load can be achieved. Retries, dead letter queues, and durable message storage are all-natural aspects of an architecture that must never lose critical information, and are all naturally supported by an asynchronous architecture.

Finally, automated testing and chaos engineering help to make the system fault tolerant by actively verifying the system's behavior in the presence of failure scenarios. Developers can find weaknesses before they become an issue in production by performing unit tests, integration tests and simulated fault injection. More specifically, chaos engineering goes even further to purposely break live areas to test the strength of their recovery mechanisms. Organizations can meet these requirements by continuously testing and improving fault tolerance through feedback loops and adding such features into their applications.

2.1.3 The Role of Load Balancing in System Reliability

Load balancing helps to improve reliability of the system by diverting the workloads to several computing resources like servers, clusters, network links, or CPUs so that no part overburdens. In the highly available and fault tolerant architecture, load balancers are intelligent traffic directors that distribute the requests evenly while gaining the fastest response times and maximum resource utilization. Load balancing alleviates the effects of resource saturation and bottlenecks in services by maintaining the smooth operating conditions while under high demand from users or when there is a sudden surge of traffic (King, 2010).

Load balancing is at a basic level a means of increasing system resilience by rerouting traffic in real time in the case of a service node becoming unresponsive or failing. Most modern load balancers are equipped with health checks and heartbeat mechanisms that facilitates the automatic detection of a failure and removal of unhealthy instances from the traffic pool. This mechanism allows the system to continue delivering service to the users without stopping the service to the users in case of partial outage or maintenance window. The very characteristic which is essential in mission critical environments where uptime is a key performance indicator.

In load balancing, there are many ways to do it, such as round robin, least connection, and IP hash-based routing, which are applied to different behaviors of applications and different patterns of traffic. For example, round robin evenly distributes the requests, least connections assigns requests to the server with the least number of active connections. In addition to L4 load balancers, application layer (Layer 7) inspection is used by advanced load balancers in making routing decisions based on content, headers or session state. It enables intelligent traffic steering which directs traffic to the most appropriate instance under some logic.

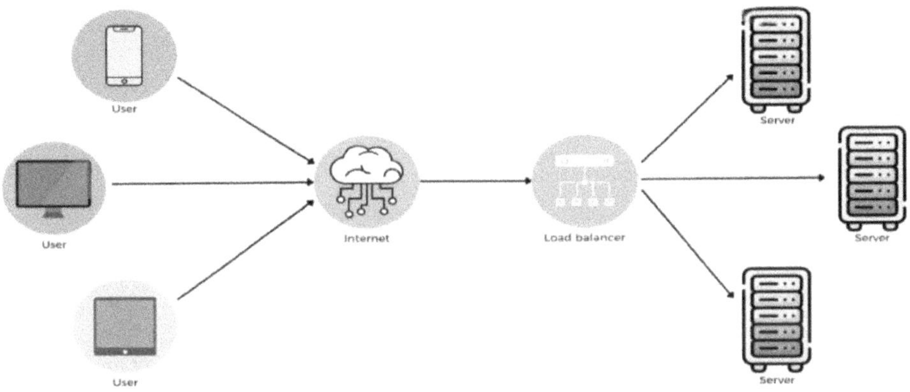

Figure 2.2: Load Balancer Architecture for Distributed Web Traffic.

Source: (Pramatarov, 2024)

Global and local load balancing work hand in hand in the distributed systems and cloud environments. They deal with traffic going across geographically spread out data centers or cloud regions for higher redundancy, disaster recovery. On the other hand, local load balancers are used to optimize data center or cluster internal traffic flow. The second part of that combination makes sure that systems can quickly adapt to regional outages, service disruptions or latency sensitive routing needs, thus results in higher system availability.

Auto scaling mechanism is also integrated with load balancing to further increase reliability. Depending on the load variations, systems can scale out or scale in the resources dynamically based on real time demand and the load balancer would automatically adjust its routing to adjust to the new resource landscape. It provides stability in case of unpreventable workloads and uses resources efficiently. The combination of containing used data with monitoring and observability, and observability, creates the opportunity for proactive intervention before performance degradation is user-visible.

In general, a load balancing is more than just a performance enhancing tool, it is a critical component of any reliable and resilient system. They distribute the requests, intelligently, across the existing systems, reroute around failures, allow for scaling activities, respond in real time to changes in traffic conditions to be available all the time and give you a consistent user experience. Dynamic control layer in modern cloud and microservices architectures for load balancing is an important means of fault tolerance and high availability of digital services.

2.2 Understanding Service Level Objectives (SLOs), SLIs, and SLAs

Re encountering the business importance of reliability and performance in a modern digital infrastructure – no matter if it's a cloud native platform, a large-scale web application or a software service is certainly a good point. Organizations adopt well defined frameworks to monitor and measure service quality to ensure that systems deliver what the users expect in a consistent manner. Service Level Indicators (SLIs), Service Level Objectives (SLOs) and Service Level Agreements (SLAs) are among the most commonly used of these (Philosophie, 2015).

Service Level Indicators (SLIs), Service Level Objectives (SLOs) and Service Level Agreements (SLAs) are the fundamental aspects of using system performance to hit business goals and user expectations. Such concepts give a structured and quantifiable way to get reliability by substituting intangible terms such as 'uptime' or 'responsiveness', for measurable metrics and performance targets, and formal commitments. SLIs are used to identify the things that should be measured to determine service quality, SLOs are used to define the acceptable thresholds for those indicators, and SLAs are used to formalize the agreed upon service levels between providers and stakeholders. Together, they constitute a strategic framework for the engineering teams in maintaining the balance of operationally stable and the innovation.

For the purpose of building resilient systems that meet their internal performance standards and retain trust from external users, it is imperative to understand the relationship between SLIs, SLOs, and SLAs. These also act as key tools to help in making a decision be it to trigger rollback, scheduling of maintenance or prioritization of development work on reliability targets. In systems of all scales and users who expect more than ever, these service metrics are even more critical. Instead of being reactive, they empower organizations to be proactive and ensure quality of service is integrated into the development lifecycle rather than as an afterthought.

2.2.1 Defining Service Level Indicators (SLIs)

Service Level Indicators or SLIs are well defined, quantifiable metrics that help us measure the performance, reliability and in general, the user experience of a system. It is the fundamental building blocks for service reliability management, the fundamental building blocks for providing a clear, data driven lens on system behavior which they provide. SLIs are chosen to faithfully record the technical qualities that matter the most to end users and stakeholders. Some of these indicators may be latency (how quickly the system responds to requests), availability (how much of the time the system is running), error rates (how often requests fail or return invalid responses), throughput (how much requests are handled in a given time), system response under load.

Defining SLIs is not just to monitor the health of system but also to reach a common understanding between engineering, operations and business teams about what makes service good. However, organizations can use these SLIs to identify and track meaningful SLIs and based on which they'll be making decisions for operational priorities, upgrades, and capacity planning. Most importantly, SLIs are different from other low-level system metrics such as CPU usage or disk I/O since they are focused on user facing metrics which directly affect service quality.

The most challenging part of choosing an SLI is the fact that the team need to find the right one with respect to the business requirements and the user journey. For example, a video streaming platform may consider buffering time and stream start up delay as one of the key SLIs and a financial services app may care mostly about the transaction completion time and error rates. A badly optioned SLI will mislead user experience, focus the User experience, or even divert attention away from the quality of reliability required. For this reason, SLIs affect the accuracy and relevance of SLOs and SLAs, which are derived from them.

Apart from helping us measure the system performance, SLIs also have an integral role to play in incident response, root cause analysis and postmortem evaluations. Teams can continuously monitor SLIs and begin to detect anomalies at an early stage, trigger automatic response or escalate issues before they impact users to a great extent. Additionally, they make it possible for smarter alerting systems with less noise and engineers with the ability to concentrate on truly impactful incidents.

While SLIs are used for monitoring, they provide an essential tool in the Site Reliability Engineering (SRE) discipline to trade between system reliability and development velocity in principled ways. SLIs quantify the user experience aspect of reliability, which is how much reliability the users have experienced, and is sufficient data to enforce reliability budgets and make decisions on when to prioritize on stability versus feature work.

In general, SLIs enable the technical and business goals of a team to be associated with the user expectations of the system. Well defined and monitored, they are the indispensable asset of engineering and business functions' service excellence and continuous improvement.

Setting and Measuring Service Level Objectives (SLOs)

Service Level Objectives (SLOs) are an important practice in Site Reliability Engineering (SRE) and system performance management in general, and are a bridge between technical operations of service and business goals. A carefully defined target, that a system or service should meet over a spanned time window, is known as SLO or Service Level Objectives in terms of availability, latency, throughput, or any other performance metrics. Service Level Indicators (SLIs) are quantifiable metrics that represent the user experience and these objectives are derived from them. Together, they ensure that a service is maintained as reliable, performant, and meeting user expectations and business needs (Ukis, 2022).

The art of setting effective SLO starts with a good understanding of the behavior of the users and the capabilities of the system. For example, engineers and product teams need to determine which elements of the service are most fundamental to user satisfaction, whether the time it takes for a page to load, or the rate of successful API responses. It is used to set realistic, but ambitious, objectives. An example SLO might be that a service should be able to respond to 95% of user requests in less than 200ms or that uptime should be at least 99.99% per month. Such goals are clear goals of performance and a small margin for inevitable imperfections.

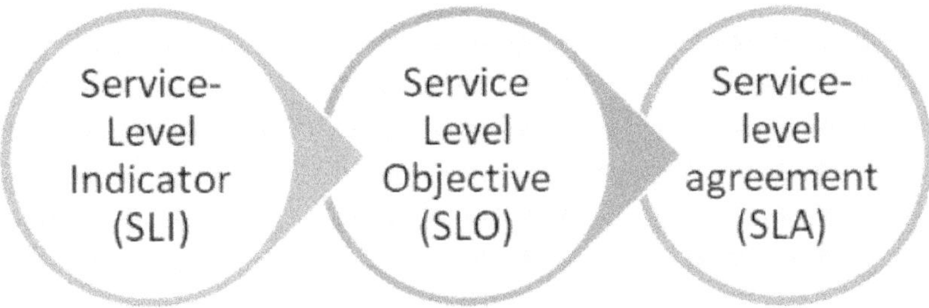

Figure 2.3: SLI, SLO, and SLA Framework in SRE for Managing Service Reliability.

Source: (A. Sharma, 2022)

An error budget is an essential part of SLOs because one of their roles is quantifying acceptable failure. The wiggle room for failure—defined by an error budget—is the buffer a service can allowance of unreliability before a certain timeframe passes. An SLO might require 99.9% availability over a 30 day month and the error budget would represent around 43 minutes of downtime. What is so powerful about this concept is that it brings the measured, controlled nature of balancing innovation and reliability to the technology product boss. The remaining error budget can be used by teams to safely roll out new features or changes. But if the budget runs out, reliability has to prevail, until the system stabilizes, and there will be no further deployments.

Through observability tools like logging platform, metrics dashboards and tracing systems, the SLOs are measured by continuous tracking of SLIs. They are tools that collect telemetry data in real time and help understand how the service is doing in terms of meeting the defined objectives. Periods of evaluation are most often over rolling time windows such as daily, weekly or monthly and this periodic analysis is used to determine trends, anomalies or chronic underperformance. Teams can be proactively watching for SLO compliance and fixing problems before they happen, spend resources correctly, and focus their work on things that directly improve reliability.

SLOs provide an engineering approach to a standardized language that allows technical teams to communicate with business stakeholders. They help to provide clarity in service performance expectations, reduce ambiguity in accountability and help to support transparency of reliability management. The alignment of engineering, product, and executive leadership results in more trust between engineering and product leaders and the executive leadership as decisions are based on measurable outcomes as opposed to subjective opinions or inconsistent data points. Additionally, SLO reviews are regular so that organizations can adjust their targets for changing user needs, new technologies, or market dynamics.

Continuous improvement is also driven by SLOs. It can also suggest that an SLO is not being met for reasons other than actual customer demand. In contrast, chronic failure to satisfy an SLO is actually an indication that the deeper problems solving them need to be delved into, such as an inefficient architecture, technical debt, or horrible incident response practices. Therefore, SLOs are not the static benchmarks but are dynamic tools for tuning and improving the system performance. They set a culture for making data-driven decisions and increase the maturity of operation in the organization.

To conclude, resilient system design and service delivery excellence requires the setting and measuring of Service Level Objectives. SLOs make teams manage risk intelligently and align system behavior with user expectations on the basis of translating abstract performance goals into concrete, measurable metrics. They enable continuous feedback loops, enhance the transparency, and keep a fine line in innovating and keeping the operational stable. In systems of increasing complexity and scale, the use of SLOs becomes more and more disciplined, and is more and more important to ensure long term service reliability and the customer satisfaction.

2.2.3 Service Level Agreements (SLAs) and Their Business Impact

Service Level Agreements (SLAs) are agreements that clearly define the provider's and its customer's level of service performance. Usually, there are usually measurable metrics like availability, latency, throughput, and so on. However, where internal benchmarks like Service Level Objectives (SLOs) only suffer from reputational ramifications from breaches, SLAs are legally binding and can have monetary or reputational costs if broken. In the SRE context, SLAs are external-facing guarantees built on top of internal SLO targets and operational indicators (SLIs). They have a function that is to set expectations for service delivery and ensure accountability (Tamer et al., 2013).

Technically speaking, SLA definitions are essential tools in transitioning customer and business stakeholder relationships with highly technical performance measures into simpler ones that are commonly understood and trusted. This enables transparency and strengthen the provider customer relationship by eliminating ambiguity around what is acceptable service performance. An example of an SLA would be 99.9% availability for a web application per month, and if the application cannot meet this threshold any failure might result in financial penalties, credits, or other compensation that is agreed upon.

SLA has its strategic weight from a business perspective. During negotiations, they affect contract adjustments, vendor management, service charging and customer satisfaction. SLA is used by businesses to measure the reliability of critical services and determine the risk. Additionally, they can function as baseline performance for organizations to measure if they are being met by a service provider. Consistent SLA meeting for service provider is very important in terms of credibility and customer loyalty whereas the repeated violation may lead to churn and damage of the service provider's reputation.

In SRE practices, engineers must ensure that SLAs are honest. The balancing act can often be quite fine; too much of overpromising strains the systems and the engineering teams; too little will deter the customers and limit growth. Setting SLAs to be slightly below SLOs does not impacted services, because it allows SRE teams to have a 'buffer zone' to absorb system anomalies or unexpected loads without impacting the business. Here engineering efforts are aligned to the long term business resilience of the organization through this approach.

Additionally, SLAs dictate how to make investments and dictate scale of infrastructure. When service performance falls under the regulated, contractual obligations, organizations more often invest in robust monitoring, redundancy, failovers, and other capacity planning to avoid breaches. In this way, SLAs get reinforced across the organization, a culture of reliability and operational excellence, both in terms of technical priorities as well as executive level decisions.

In the end, SLAs are not technical metrics, but the service provider's promise to deliver a consistent value. SLAs become strategic instruments that dictate a company's service quality, customer experience and market competitiveness when developed collaboratively between SRE team and business stakeholders. A good SLA is a cornerstone of reliability as it shows the organization's commitment to making excellence and to delivering under pressure.

2.3 Capacity Planning and Load Balancing

There are two fundamental pillars in the context of capacity planning and load balancing in the design and operation of reliable and scalable systems. These are the strategies on which system performance should be maintained as Real time, system continuity should be maintained as Real time, to maintain Business Continuity and system performance maximally at the same time. Capacity planning is the act of anticipating the number of resources needed by a system to process varying levels of demand over time. This makes sure that applications have always sufficient computing power, storage, memory, bandwidth, and so on to run their jobs without being under or over provisioned. It is a forward looking process in which data from past trends, anticipated business growth and seasonal changes are used to plan for future conditions (Arun Kejariwal, 2017).

Effective capacity planning prevents these common pitfalls of system outages which occur when resources are exhausted or idle infrastructure becomes wasted for unnecessary costs. Site reliability engineers (SREs) can use real time observability and analytics tools to monitor such metrics as CPU utilization, memory saturation, request latency, disk I/O. This helps in making decisions about the way ahead – of scaling and resource allocation. Teams can have the foresight to foresee actual hardware upgrades, as well as cloud resource adjustments, by using forecasting models and simulation techniques. This is especially crucial in the current digital services where even the slightest performance slippage may cause distrust or revenue loss in the services.

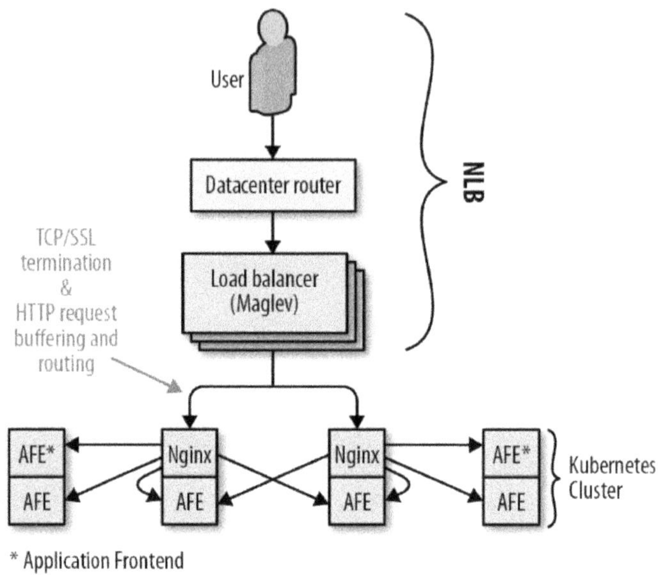

Figure 2.4: Load Balancing Architecture Using Maglev and Kubernetes Cluster.

Source: (Cooper Bethea, Gráinne Sheerin, Jennifer Mace, 2019)

Since workloads change over time and systems cannot always be well provisioned, they have to be able to distribute traffic over time efficiently. For this, load balancing comes into picture. The process of load balancing is accomplishing an even distribution of the traffic across multiple servers or services to avoid that any one part would become overwhelmed. It makes the applications more responsive, prevents the service degradation, and increases fault tolerance. There are various ways in which load balancing can be achieved – through software based solutions, hardware appliances or managed cloud service, which are not all equally difficult to manage, efficient and affordable.

Load balancing in modern cloud environment is more than just distribution. Often the intelligent traffic routing is based on server health, user location, response time, and availability zones. It provides more performance, but most importantly it's also resilient. For instance, traffic is automatically rerouted to healthy instances at other servers, so that service availability is not dropped in case of a server or a region failure. Achieving high availability of distributed systems requires support for this kind of failover. To further enhance fault tolerance, redundancy is introduced through multiple replicas of services or geo distributed infrastructure such that user requests are still served even when some part of the system fails.

In globally distributed systems, where users access services from various parts of the world, load balancing is one of the more advanced application. In this case, the user requests will have to be routed to the nearest or most responsive data center to minimize latency and improve performance. Service seamless is achieved through these strategies through DNS level routing, Anycast networking and Traffic steering algorithms. Besides enhancing user experience, it is a safety measure during regional outages balancing reliability with speed.

There is monitoring in capacity planning and load balancing. Early detection of issues or proactive improvement of systems is nearly impossible when our visibility for how they are performing in real time is abstracted in response time values. The tools a site reliability engineers use include Prometheus, Grafana, Datadog and cloud native monitoring platforms to track performance indicators, error rates, traffic distribution, etc. There are alert mechanisms and automated response to mitigate issues before they affect the user. Through an observability driven approach, this is a reliability engineering, instead reactive firefighting.

The applications of capacity planning and load balancing are shown at the end of the article across a wide variety of industries. In fact, these strategies are used heavily during peak shopping seasons, such as Black Friday or holiday sales to ensure uninterrupted service on the part of online retail platforms. They make streaming service possible by delivering consistent content through various geographies by ensuring the same content to the customers. Furthermore, these principles also help in avoiding disruptions to the critical services such as authentication, payroll, reporting and others that have been built as internal enterprise applications. Such systems bear effectiveness in maintaining an optimal balance between resource availability, and intelligent traffic management.

In a nutshell, capacity planning and load balancing are not a onetime activity, but on going process and need to change according to changing business needs and technical challenges. All together, these are the components that make up system reliability, scalability and performance. These strategies will become even more important the more and more digital services become complex and large. Organizations that set up plans, and have dynamic traffic management in place, are more capable of providing constant high-quality service experiences to their customers while minimizing cost and also reducing risk.

Forecasting Demand to Avoid Over-Provisioning

With the digital age comes the need to predict system demand in order to achieve the service efficiency and operational cost effectiveness. As more and more organizations are turning to cloud native infrastructures, a scalable platform and some distributed systems to serve shelf edge users, the need to forecast demand accurately and timely becomes more and more critical. The initial solution to avoiding downtime may seem like overprovisioning by allocating more resources than are necessary, however this has its own set of costs and inefficiencies. This leads to an underutilization of computing power, a higher infrastructure expenses, more energy and consequently a negative environmental impact. Furthermore, the idle resource management can also complicate the operations and diminish the overall system agility (Publishing, 2025).

Working well with demand forecasting gives the organizations the freedom to make a strategic balance: supply as much as required to cope with fluctuations of user loads, but not too much resources which are unused. During high traffic events such as the product launch, seasonal sales or promotional campaigns, the important thing is this balance as a sudden traffic spike can stress the system. By using historical usage data, the predictive analytics and machine learning models, the teams will be able to forecast the future demands better and react to the future resource allocations accordingly. It not only guarantees the performance stability and reliability, but also optimizes resources utilization and minimizes operational waste, which in turn makes it more suitable for the matching of technology investment with organizational business needs.

- **Why Demand Forecasting Matters**

Usage of every digital system, from e-commerce to streaming, and from enterprise to enterprise, is fluctuating constantly. The reason behind these fluctuations can be a predictable one (like time of day, make money with foreign companies seasonality) or unpredictable ones (like viral content, technical issues). By forecasting demand, organizations can prepare their infrastructure to deal with the expected as well as the unexpected and not burden their budget.

- **Data as the Foundation for Forecasting**

In order to forecast, steps would be taken in analyzing historical usage data - CPU and memory utilization, network traffic, application requests to name a few - in order to predict. With this information, patterns and trends can be identified and anticipated future behavior. Both of these include the application of predictive analytics and machine learning methods allowing the forecasting methods to be more accurate and adaptable over time.

- **Risks of Over-Provisioning**

As far as over pro visioning goes, this may sound like a safe strategy, but it really isn't, because it comes at a cost of unnecessary cost and environmental impact. Servers sitting idle still consume electricity and maintenance resources. Unused capacity is money that is lost in cloud based systems because the services are billed by usage. Also, it makes it complex in monitoring and managing surplus resources.

- **Finding the Right Balance**

Demand forecasting has nothing to do with being able to prevent wasting resources, actually, it's about designing systems that are responsive and scalable. Accurate forecasting with auto scaling mechanisms enables real time growth or shrink of the systems. The presence of this provides businesses with the ability to respond to a surge in demand that they did not expect and avoid constant over-provisioning.

- **Adapting to Changing Needs**

Demand forecasting is an ongoing activity, and it is not a one-time thing. New features need to come out, and as user behavior changes the forecasting model needs to change. A continuous evaluation helps businesses remain responsive to the demands while keeping operational costs low and still optimize its performance.

2.3.2 Load Balancing Algorithms and Techniques

Load balancing is a key technique in distributed systems and cloud computing as it brings the workloads evenly apportioned across various servers or resources for better response time, reliability, and preventing the system malfunctioning. Implementation of different algorithms and techniques for load balancing through deciding traffic allocation among the available resources at heart of successful load balancing. To achieve these goals, these algorithms tend to maximize throughput, reduce latency and maintain a fair resource utilization which are considered as the foundation of high performing and resilient infrastructure (Jafarnejad Ghomi et al., 2017).

Depending on the environment, there are different types of load balancing algorithms. Round Robin is one of the most common, it sequentially distributes requests to the server pool repeating the loop at the end. However, this approach is good in environments where servers are equally capable and performant. Yet, Weighted Round Robin is more common when servers have different capabilities, as it is able to distribute more powerful servers a larger share of requests by assigning higher weights.

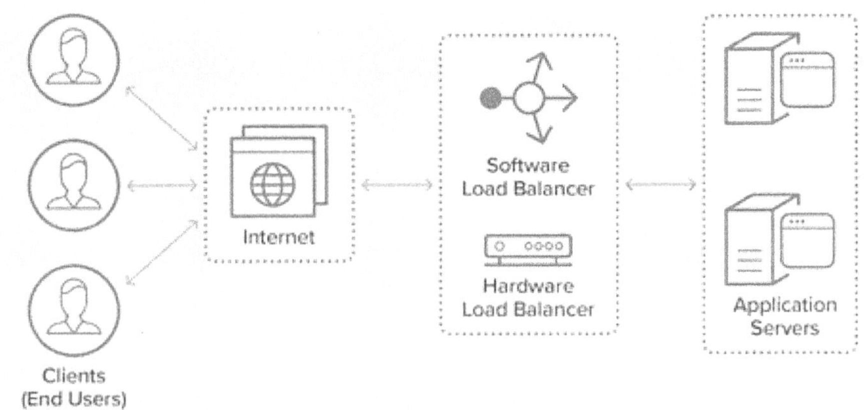

Figure 2.5: Architecture of Load Balancing Between Clients and Application Servers.

Source: (Vertisystem, 2023)

Least Connections is another widely used technique, the incoming traffic is directed to the server with the least number of active connections. In the case where session lengths are unpredictable, this method is appropriate where no one server is overburdened. Another IP Hashing technique is used to route the request to a specific server based on the client IP address to make the same client always go to the same server for session persistence.

Dynamic load balancing is more advanced and decisions are made based on real time metrics such as, CPU utilization, memory usage, etc, and more importantly server health status. These algorithms can also be switched in real time that is, they can change strategies in response to unstable traffic, and system anomalies while still providing optimal performance. Another function that load balancers may perform is that they may also do health checks and automatically redirect traffic away from failing or loaded nodes, which will increase the fault tolerance and system uptime.

Loading balancing is a crucial concept in the cloud native environments where technologies such as Kubernetes that primarily depends on the load balancing to handle pods and services acros distributed clusters. Thanks to tools such as Envoy, HAP Roxy, Nginx and cloud native services from AWS, Azure and Google Cloud, load balancing is as highly configurable as traffic management can be and it becomes possible for DevOps and SRE teams to dink around with traffic management based on application needs.

In general, the system reliability, efficiency, and scalability are all dependent on understanding and choosing the appropriate load balancing algorithm. These techniques are indispensable in ensuring that digital application becomes increasingly complex and user expectation rises, there is no hitch in the user experience.

2.4 Redundancy and Disaster Recovery Strategies

In modern digital infrastructures, it is vital to make sure the systems are reliable and business continuous. These objectives can be achieved with the aid of redundancy and disaster recovery strategies. Redundancy refers to the replication of critical components or functions of the said system to ensure reliability and availability. However, disaster recovery is the structured approach that services us when an event that was unplanned such as cyber-attack, natural disaster or system failure happens (Birman, 2005).

Implementation of redundancy can be performed at different levels of hardware, software, network, and data storage. For example, servers are deployed across multiple availability zones in order to mitigate failure impact at a localized level. redundant paths and failed routing make it possible to keep with connectivity even when the primary link goes down. Similarly, running multiple data centers with synchronized systems makes it easy for a business to switch over to other operations in cases of outages.

For disaster recovery, strategies as to prepare for what will not be predictable, and minimize the downtime. Typically, this would include regular data backups, secondary or tertiary environments for fail over and define clear Recovery Time Objectives (RTOs) and Recovery Point Objectives (RPOs). RTO is the time that systems need to be recovered and RPO is the maximum amount of time that can be lost in data. They jointly plan and resource allocation for efficient disaster recovery.

Disaster recovery (DR) and recovery of business services is no longer done individually for each application, but is instead being enabled more now than ever by technologies such as cloud DR (DRaaS), virtualization and container orchestration tools such as Kubernetes. These tools facilitate rapid replication, etc. and automates recovery process that eliminate the need for manual interventions and the recovery time.

Resilience Engineering strategies should not be composed of separate efforts to increase redundancy and disaster recovery. Regular testing, updates, and audits of them are required because systems and business needs change. By proactively investing in these strategies, such leaks could help organizations to reduce risks for data loss, maintain service continuity, and safeguard the paramount reputation even during catastrophic events.

The effectiveness of these redundancy and disaster recovery strategies ultimately hinges on how closely they conform to the organization's priorities for redundancy and disaster recovery; the level to which the organization is willing to tolerate risk; and how compliance is required.

They operate when executed well to make sure that even in the face of failure, the business continues to perform without major interruption.

2.4.1 Implementing Multi-Region and Multi-Cloud Strategies

As cloud computing and Site Reliability Engineering (SRE) become increasingly fast paced, multi-region and multi cloud strategy has become a critical implementation for achieving very high levels of system availability, resilience, and performance. The goal of these strategies is to proactively prevent risks for having a single point of failure or regional outage and cloud vendor dependency by becoming intelligent about where we workloads and services live. However, through the architectural design, this places nothing in the way of uninterrupted service continuity, low-latency and operational agility in the presence of large scale disruption or infrastructure failure, as classical large and distributed systems commonly experience (Kleppmann, 2017).

A multi-region strategy is deploying infrastructure and application component across multiple physical or virtual regions in a single cloud provider's global footprint. Thus this geographic dispersion affords greater fault isolation and disaster recovery, so that a failure or capacity暄 neutrality of one region does not prevent a fault. Also, it lets organizations serve user requests from a nearest regional data center thereby reducing latency and improving end user experience. In particular, this model is important for global scale applications in which both performance and availability are strongly expected.

In addition to this approach, a multi cloud strategy involves the use of multiple cloud service providers (CSPs) such as Amazon Web Services, Microsoft Azure, Google Cloud Platform (GCP), etc. in order to spread out the operational risk and avail maximum flexibility. Organisations instead use the unique service offering and pricing models that the providers offer to ensure that they don't become locked into a vendor, attain regulatory compliance across jurisdictions and deploy the better of breed vendor for particular work loads. Furthermore, this approach permits for strategic redundancy to operate essential services even in the event that one provider has a devastating outage or outage.

They need a set of advanced tools and processes to be effective in operationalizing multi-region and multi-cloud. It also includes global load balancing, anycast DNS, traffic management policies, all of which are extremely important to global and across the cloud consistency. IaC tools like Terraform or Pulumi help ensure a consistent environment provisioning way, and observability solutions, such as Prometheus, Grafana and distributed tracing systems help to see the health, the latency and the reliability of the system on all environments.

Additionally, replication policies and secure data pipelines along with resilient storage are must for data consistency, synchronization and compliance at regional as well as cloud level. Such is the complexity of this problem that it is complex enough to require a skilled DevOps and SRE staff to design, deploy, and maintain such distributed systems in line with the business goals and service level objectives.

It's a complicated strategy, and the operation involved is complicated as well, but the payback is huge. That is why these strategies are so robust in terms of fault tolerance, building customer trust and business continuity with the only way that any business can provide 24/7, globally connected service today. Currently, demand for scalable, resilient, low latency services is increasing, causing these techniques to be used in modern cloud native system design as an intrinsic part.

2.4.2 Active-Active vs. Active-Passive Architectures

Today in the digital world of interconnectedness, the availability of service for uninterrupted service availability is critical, and the Active-Active and Active-Passive architectures are the primary building blocks for system resilience and fault tolerance. Given the importance of such architectural strategies to designing systems with good resilience to failures, scalability during load surges and SLA guarantees, these are one of the more intriguing questions to answer. They are deployed in enterprise IT infrastructure, cloud-natives platform, and globally distributed services to maintain business continuous and without disruption the user experience (Lydia Parziale, Guillaume Lasmayous, Manoj S Pattabhiraman, Karen Reed, 2014).

In an Active-Active architecture, there are multiple nodes or data centers that are up and running as well as providing live traffic. In the participatory node, the application workloads are handled by all participating nodes collaboratively, shared resources and consistent performance across geographies. It is this type of architecture that gives built in redundancy, if a node fails or becomes unreachable the others will automatically pick up the load without causing any noticeable service disruption. This setup has an important advantage of capability to improve the performance since the user requests are routed to the closest or least congested node to reduce the latency and improve the user experience. Furthermore, this architecture supports real time data replication and distributed processing that are most suitable for industries like e-commerce, financial services and media streaming where even less than milliseconds can make a big difference in the level of user's satisfaction or a transaction success.

However, the complexity of Active-Active systems is high. All instances must be in sync and avoid conflict in data, so it is a sophisticated traffic management policy, robust consistency protocol, and real time data synchronization mechanism. Usually, besides global load balancers, using distributed databases with conflict-free replicated data types (CRDTs) and consensus algorithms (e.g., Raft or Paxos) are used to guarantee system integrity. At the same time, the operational overhead and financial expense of operating different live instances need to be considered carefully because it comes with redundant infrastructure, continuous monitoring, and rigorous testing.

The Active-Passive architecture is much simpler and cheaper model where there is just a single active node doing the production workload while the one or more passive nodes are on standby. Under normal operation these standby systems do not process live traffic and keep up with the active node in order to be ready for failover on a fault. The system can have automatic or manual failover mechanisms which depend on system configuration and recovery time objective (RTO). It is widely used as a backup system, disaster recovery environment, and applications that require little uptime.

The Active-Passive setup has an advantage of reduced complexity and operational cost. The infrastructure is cheaper since only the active node is used during normal operations. Furthermore, switching roles between active and passive nodes can be done to perform maintenance tasks such as upgrades or patches without compromising availability. Despite this architecture there is a brief service interruption for failover, which may be undesirable for real time systems. In addition, as the passive node is not routinely load tested under production conditions, there is a possibility that problems will arise as it moves to active status in emergency.

Both architecture implementation decisions should be based on specific operational goals of the organization, and budget constraints and the expectations of the users. Furthermore, high frequency trading platforms, online banking services and healthcare systems are also very much benefiting from zero downtime of Active Active set ups. On the other hand, business applications that do not require as much uptime as critical applications like Active-Active may be more suitable for Active-Passive.

Both architectures are crucial to the reliability toolkit from an SRE perspective. Active-Active fits with SRE's error budgets, observability, and scalability mantra because it supports proactive service health monitoring and automatic recovery. Active Pasive on the other hand supports simplified incident response, better resource allocation, enabled controlled failover testing. In addition, the architecture choice determines how SREs will define and measure the key reliability metrics such as service level indicators (SLIs), and service level objectives (SLOs), which in turn defines the ways by which availability, latency, and recovery processes are tracked.

Finally, it is represented that the design and management of resilient digital systems heavily depend on both Active Active and Active Passive architectures. Active-Active maximizes availability and performance through real time redundancy and is somewhat more expensive than Active Passive, but Active Passive provides a simpler and often the less expensive solution for maintaining service continuity. The selection of right model involves tradeoffs between complexity, cost and availability level needed, and balancing that correctly is indeed an art of modern infrastructure design and site reliability engineering.

2.4.3 Designing a Disaster Recovery (DR) Plan

The majority of business continuity approaches involve creating a Disaster Recovery (DR) plan for business continuity, operational resilience, etc. DR plan that is laid out in a well-structured manner is an outline of the procedures and protocols to be followed in case the system goes off due to outages caused by natural disasters, cyber-attacks, hardware failure or any other such events. The goal is to reduce downtime and data loss to reduce the amount of time the organization remains down and unable to operate normally. A comprehensive, scalable, and tailored plan needs to be made for the organization and its infrastructure (Aljuhani, 2017).

The identification and classification of critical systems, applications and data that are essential to do business is the first step in creating a DR plan. It is achieved through a thorough risk assessment and business impact analysis (BIA) to determine priorities in terms of recovery efforts, as the consequences of disruption will vary depending on the significance of the parties involved. RTOs and RPOs should be defined for each critical asset within the plan so that there are measurable targets for system restoration and data recovery. These objectives are used in the selection of suitable backup solutions; failover mechanisms; and data replication strategies.

Then, the technology and infrastructure that should aid in recovery efforts should be included in the plan. It includes offsite backups, a geographically spread out data centers, and automated failover systems that take over the operations in a standby environment during an outage. Disaster recovery solutions have become more popular lately because of their cost effectiveness as well as their scalability. With the help of these tools, the recovery time can be drastically reduced and flexibility can be applied to managing various workloads. To keep the DR plan up to date, it is important to update regularly and to have it under version control.

Finally, a DR plan should be tested and updated on a regular basis to be effective. Due to periodic disaster recovery drills and simulations, weaknesses in the plan can be identified and the team can be prepared. It should be clear and accessible, and everyone involved in the disaster scenario, including IT people, business leaders and external vendors, should know what they are doing. Continuous training and improvement not only strengths the plan's reliability, but also reinforces the overall resilience posture of the organization. A DR plan well executed serves as insurance protecting trust, data and service continuity in a digital world that is becoming more and more unpredictable.

Figure 2.6: Disaster recovery strategies.

Source: (INFOSECTRAIN, 2022)

1. Data Center Replication

The thrust of this approach is continuous replication of data among multiple data centers. If there is a failure at one location, systems can fail over to a replica right away without ever missing a beat. It reduces data loss and downtime, which makes it one of the basic strategies for redundancy.

2. Disaster Recovery as a Service (DRaaS)

Third party providers are responsible for disaster recovery processes in DRaaS, which is a cloud-based service. This is especially useful for small to medium size organization which lack of in-house resources might not allow them to set up and maintain complex DR systems. This is an on-demand scalability that reduces the infrastructure prices and rapid recovery options.

3. High Availability (HA) Systems

These are the systems intended for continuous operation with no interruption. Hardware and software multiple layers of redundancy is a part of a HA setup to prevent single points of failure. For mission critical applications, any downtime is too much and this strategy is critical.

4. Cloud-Based Disaster Recovery

The method is to back up data and applications to cloud environments, such as cloud storage and computing resources that are geographically distributed. When a disaster occurs, cloud services have the advantage of deploying in a flexible and timely manner without relying on the physical infrastructure.

5. Continuous Data Protection (CDP)

CDP captures the data changes in real time and saves them. Unlike traditional backups, that are performed at scheduled intervals, CDP records all changes instantly. This permits restoration of the systems to any point in time, thereby offering considerable reduction in data loss during recovery.

6. Data Backup and Restoration

Regular backups are still very much on the list of disaster recovery fundamentals. This strategy is by making copies of data at regular intervals and keeping them safe. Restoration is the process of getting lost or corrupted files, systems or applications back using backed up data after a failure.

7. Disaster Recovery Sites

These are secondary sites which are located in different geographical regions of the same site with the same data and operations mirrored to them. The organization can shift its operations to the recovery site in the event of a disaster in the primary site with minimal downtime. And these can be 'hot' (fully operational and updated in real time), 'warm' (partially ready) or 'cold' (basic setup which will take more time to become operational).

Overall, the various layers of strategies that organizations must take to become completely prepared for the calamity. Replication, latency eliminated cloud-based recovery, traditional backup and redundant systems are all strategies that businesses can adopt that allow them to continue to operate, protect business critical data and quickly recover from unexpected disruptions.

2.5 Traffic Management and Load Shedding

Traffic management and load shedding are major techniques used in modern digital infrastructures to keep system in balanced, stable state while ensuring enough traffic experience during the peak of loads or unexpected demand surge. In both a global scale in applications and a dynamic user demand, the design to be able to intelligently route traffic and shed non critical loads is a cornerstone of resilient system design. The traffic management will make sure that multiple servers or regions will get requests efficiently so that performance will be optimized and latency and bottleneck will be reduced (Chakravarty & Chiang, 2018).

However, load shedding is a protective mechanism used when the system capacity is threatened. Here, it simply entails deliberately denying or de-prioritizing some types of requests (less important, or less critical traffic) in order to free up resources for critical operations. This strategy will make sure that, even under extreme load conditions, the high priority services are still functional. For example, background analytics processes may temporarily pause in an e-commerce platform to have it run at full capacity in order to cater to a flash sale.

In order to have an effective traffic management, the following technology is mandated — a global load balancer, DNS based routing, CDN and application aware proxy. Together, these machines guide traffic of users towards different locations depending on proximity, given latency, healthy status of endpoints or latest demand dynamically. They help in using the resources optimally, Increase user satisfaction and ensure consistent uptime. Additionally, it can support A/B testing, canary deployments, and blue green releases that allow organizations to roll changes without risking much.

Traffic management and load shedding are a strong defense to service degradation, working together. If these practices are used when designed and implemented as part of a larger Site Reliability Engineering (SRE) strategy, then systems will be responsive, efficient and fault tolerant. Organizations can deliver high quality digital experiences to all critical customers under all conditions—planned or unexpected—and do so by managing demand proactively and prioritizing critical workloads.

2.5.1 Managing High-Traffic Scenarios Effectively

In modern digital systems, high traffic scenarios are turned into an essential requirement of performance, availability, and user satisfaction. However, with applications and services generating unpredictable traffic spikes – marketing campaigns, product launches, or seasonal sales for instance – organizations must have staggering methods of making sure that their systems can take in that spike without compromising user experience. High traffic management is successful when these predictive planning, real time traffic routing, auto scaling, caching, intelligent queuing systems are mixed together (Wright, 2022).

In general, these are some important strategies and concepts typically involved:

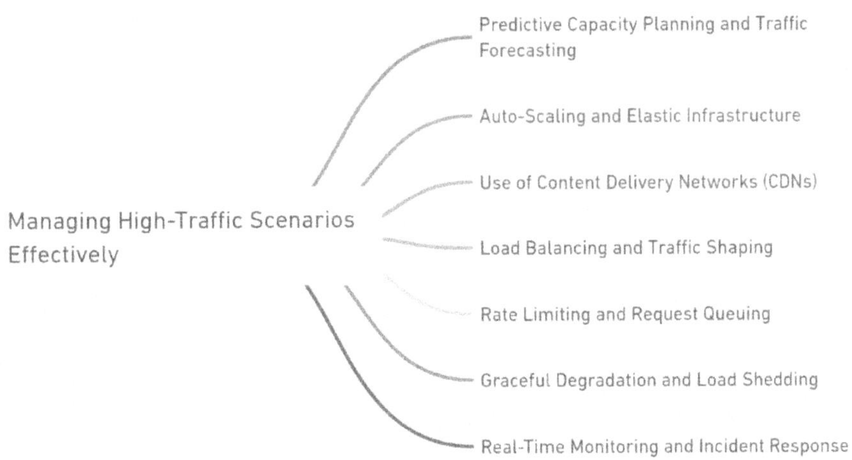

Figure 2.7: Strategies for Managing High-Traffic Scenarios in Web Systems.

Source: (Self-generated)

1. Predictive Capacity Planning and Traffic Forecasting

To plan for traffic surges, organizations are using historical traffic data, machine learning models, usage trends and so on. That is the advantage of predictive analytics, they can predict the traffic and ensure that there is enough resources available before it reaches traffic spike. It reduces slowdowns and outages risks.

2. Auto-Scaling and Elastic Infrastructure

It means that with cloud native platforms, compute, storage, and online resources (network) can scale itself automatically on the basis of real time traffic metrics. Horizontal scaling (increasing number of instances) and vertical scaling (increasing the capacity of existing instances) helps to absorb sudden bursts of demand without human intervention.

3. Use of Content Delivery Networks (CDNs)

Real time traffic metrics are a data point for automatic scaling of compute, storage, and network resources without any provisioning or scaling considered by the platform. Horizontal scaling (bump the instances up) and vertical scaling (belt out the existing ones) can be used to absorb sudden surges in demand without manual intervention.

4. Load Balancing and Traffic Shaping

In fact, intelligent load balancers distribute incoming requests to healthy backend instances based on different criteria such as round robin, least connections or even geographic location. Traffic shaping can also be advanced such that critical services are prioritized over less critical ones.

5. Rate Limiting and Request Queuing

Depending on the system, incoming requests can be prevented from overloading the service, throttling according to user type or endpoint, or queued in an orderly fashion. Rate limiting protect the backend services from abusive traffic patterns or sudden bursts, and promulgate the fairness and reliability.

6. Graceful Degradation and Load Shedding

As systems get closer to the limit, some of the non essential parts can be turned off temporarily, such as personalized recommendations or background jobs to save resources for mandatory work. It guarantees that critical tasks such as login or payment processing will be available.

7. Real-Time Monitoring and Incident Response

For managing, it's necessary to have robust observability tools to know the traffic patterns, latency, error rates in real time. There are systems for alerting and the runbooks to respond if anomalies are detected.

2.5.2 Rate Limiting and Circuit Breakers in Action

In the modern world, maintaining service reliability and performance under varying load condition is crucial in the distributed systems. Two defensive design patterns used for traffic spikes, canceled failures and to keep the system applying resilient—rate limiting and circuit breaking. If implemented properly, these mechanisms play the role for infrastructure of being intelligent and robust guardians, maintaining that services are still available and interactive, even with stress or partial outages.

The rate limiting technique restricts the number of requests that a client, user or service can send to an application within a period of time. It acts as a preventive control to avoid misuse and abuse, and unintentional overload. This is very important in APIs, authentication services, payment gateways, as well as shared resources that should degrade in case of an overload. Algorithmic models like token bucket, leaky bucket or fixed window counters are usually used to enforce rate limiting. These algorithms help in determining how many requests can be allowed and how fast the tokens (allowed action tokens) will be replenished.

The one real case of use is in the deployment of public APIs where rate limits for limiting fair usage of APIs from consumers and protection from backend infrastructure. For instance, perhaps a weather API would let a user make 1,000 requests per hour per user. After that threshold, users get HTTP 429 responses stating "Too Many Requests." Not only it will protect system integrity, but it will also increase the availability of service for all users.

In cloud environments, where services such as serverless computing or managed APIs are billed on usage, it is also useful for cost control (as rate limiting doesn't restrict throttle throttling in those service definitions). Limiting the requests in teams helps in avoiding unexpected billing spikes and keeps the model cost predictable. Furthermore, dynamic or adaptive rate limiting is a much used method to adjust limits based on the server load or in the present case the current response times, improving the system responsiveness during the peak traffic.

On the other hand, circuit breakers are meant to gracefully handle all failure issues. On a micro services architecture, any one slow or failing service can bring down everything your chain of dependent services are all digesting. Failure conditions to monitor are timeouts, increased latency, high error rates, and circuit breakers 'break the circuit' and temporarily stop requests to the problematic service. It prevents resource exhaustion and it allows the failing component time to recover or alert the engineering team to intervene.

There are three states of a circuit breaker, which are closed, open and half open. Requests normally flow in the closed state. In the event failing thresholds are hit, the circuit moves into an open state, and stops traffic to the impacted service for a fixed duration window. Once this window has passed, the circuit goes into half open state, and a small amount of test requests are allowed to decide whether the service is recovered. If so, the circuit closes, otherwise, it stays open, preventing the system from being damaged.

Circuit breakers do a good job in action in e-commerce systems during the sale events. A recommendation engine or payment service can run slow if there is a sudden demand and the circuit breaker will halt requests to that component while still functioning the main shopping cart and checkout process. It enables customers to continue to browse and buy without experiencing full system outages.

Rate limiting and circuit breakers are two key patterns in Site Reliability Engineering (SRE) that help to guarantee the service level objectives (SLOs) are routinely met. Beyond provisioning metrics, logs, and traces, it gives a control mechanism to address actions. Furthermore, these patterns can be correlated to auto scaling strategies and alerting systems so that better automation of response is put in place.

Finally, all is concluded with rate limiting and a circuit breaker which was able to prevent faults by isolating them and the load. Implementation of these relationships offers the way for organizations to construct more sufficient, oversight fault resistant, and client centered digital encounters. These mechanisms are not just helpful but indispensable as systems keep scaling in complexity and need, in order to enable growth and reliability in production environments.

Multiple Choice Questions (MCQs)

1. **What is the primary goal of designing for high availability?**

   A) Reducing code complexity
   B) Improving UI design
   C) Minimizing server costs
   D) Ensuring continuous system uptime

2. **In fault-tolerant systems, which of the following components is crucial for minimizing the impact of a failure?**

   A) Redundant components
   B) A/B testing
   C) Load testing scripts
   D) Centralized logging

3. **Which of the following best defines a Service Level Indicator (SLI)?**

   A) A backup plan for infrastructure
   B) A measure of system performance from the user's perspective
   C) A feature for reducing latency
   D) A contract between the business and a third party

4. **What is the purpose of setting a Service Level Objective (SLO)?**

   A) To define internal performance targets
   B) To estimate future sales
   C) To create user interface guidelines
   D) To improve storage performance

5. **Which of the following best describes a Service Level Agreement (SLA)?**

   A) A legally binding agreement outlining expected service levels
   B) A technical manual for software release
   C) A list of servers and configurations
   D) A data compression technique

6. **Capacity planning helps prevent which of the following?**
   A) Overuse of third-party APIs
   B) Over-provisioning or under-provisioning of resources
   C) Increased testing overhead
   D) UI inconsistencies
7. **Which load balancing algorithm distributes traffic based on the server with the least number of active connections?**
   A) Round Robin
   B) Fixed Priority
   C) Least Connections
   D) Hash-Based
8. **What is a key benefit of using a multi-region deployment strategy?**
   A) Minimizes build times
   B) Improves fault isolation and global availability
   C) Reduces the need for documentation
   D) Reduces dependency on DNS
9. **What is the key difference between Active-Active and Active-Passive architectures?**
   A) There is no difference
   B) Active-Active is for testing, Active-Passive is for production
   C) Active-Active systems share load, Active-Passive systems rely on a backup
   D) Active-Passive systems require more users

10. Which of the following is used to manage high-traffic spikes and prevent system overloads?

A) Rate limiting and load shedding
B) Feature flags
C) Caching invalidation
D) Data replication

Answers

1.	2.	3.	4.	5.	6.	7.	8.	9.	10.
D	A	B	A	A	B	C	B	C	A

CHAPTER 03

Observability, Monitoring, and Incident Management

3.1 Understanding the Three Pillars of Observability: Logging, Metrics, and Tracing

In an age where the internet is the next line after its name, and AI is everywhere, the reliability of complex systems is a key aspect—industry like oil and gas can't afford for a minimal downtime to entail an operational or financial tip. With the growing trend towards distributed and cloud native architecture, the ability of monitoring and understanding the system behavior is essential to business enterprise. With this need for visibility growing, observability has emerged as the practice of enabling teams to get actionable insights about the internal state of the systems they run, by the data they produce (Jez Humble, Joanne Molesky, 2014).

Logs, metrics, and traces are the three main observability pillars. Each of these pillars has a specific part to play in drawing out a full picture of what system behavior is. Logs store granulous, time stamped sets of events, and their main value is for logging, troubleshooting and post incident analysis. Metrics provide numerical indices of the system performance, e.g. resource use and response times that can be monitored on real time and able to be analyzed in trend through time. The traces are similar to the traces in microservice based architecture however the trace is a detailed view of the path that request took within a system indicating latencies and identifying bottlenecks.

These observability components, when used together, enable proactive system management. Instead of just dealing with failures, organizations provide an opportunity to anticipate, to understand root causes more quickly, and do so before problems impact critical operations. In the field of oil and gas, for instance, real-time decision making, predictive maintenance and system optimization are becoming of key importance to the digital transformation initiatives. Thus, observability becomes more than a technical capability and becomes a strategic enabler.

Figure 3.1: The Three Pillars of Observability

Source: (Inspirisys, 2023)

This chapter examines the way in which logs, metrics and traces make operational space more transparent and responsive. The goal is to make readers aware of the benefits the observability can bring on the issues of the system performance, reliability and using AI tools for monitoring and automation. With observability practices adopted, they position themselves to be on top of a more and more data driven and automated industry.

3.1.1 Logs: Capturing and Analyzing System Behavior

Log files are a system's history records. They frequently come in binary, plain text, or structured format and are time stamped. Text and metadata can be combined in structured logs to speed up querying. (Learning & Reserved, 2010).

In the event of a system error or malfunction, administrators initially check the system logs to determine what went wrong. Logs keep track of occurrences and can identify unusual activity, like a security lapse in the load balancers or caching systems of its infrastructure. It can also assist in providing answers to the questions of "who, what, when, and how" resources were accessed.

Benefits of Logs

Although debugging and troubleshooting are the usual uses for logs, they might offer other advantages.

- **Alert and monitoring:** To inform users before they are impacted, notifications can be created on particular logs or logging patterns. Additionally, engineers may manage event data through search and curated services because of the monitoring tools.
- **Managing resources and troubleshooting:** Administrators utilise log data to track events and patterns across systems, keep an eye on anomalies or inactivity in real time to assess system health, identify configuration or performance problems, troubleshoot data analysis, and identify the root cause of problems.
- **SIEM and regulatory compliance:** The IT department may increase efficiency and comply with regulations by automating the collection, analysis, and correlation of log data from several security systems and devices. SIEM systems oversee workflow, real-time analysis, logs, and alerts.
- **Business analytics:** When certain business objectives are met or requirements are met, log data can be mined to extract important

business information, such as hourly revenue, customer service level agreements (SLAs), and the health of company processes.

- **Marketing insights:** Digital marketers can gain insights from log analysis on the visibility, traffic, conversions, and sales of their campaigns. It might, for instance, show how bots browse its website, opening it new SEO possibilities like identifying useful and pointless information. Display both the crawled and non-crawled pages from Google. Boost decision-making and forecasting. notifications of important occurrences or trends, as well as better-track websites.

Limitations of Logs

Although logs have many advantages, there are drawbacks as well.

- **Large amount of data:** While recording every event aids in understanding the current state of the system, it also raises the requirement for data storage. Businesses must utilize a log management solution that will help them gather only the most important log data in order to make the most out of the occurrences. particularly those built on systems with a lot of microservices, where concurrency isn't reflected in logs.
- **Higher cost:** Due to the space required to keep all the data, maintaining extensive logs for an extended period of time can be costly. For instance, adding more containers or instances raises the cost of logging and storage in order to handle greater client activity.
- **Performance-related problems:** Although creating a log is easy, if the logging library does not enable dynamic sampling, excessive logging may cause the program to lag. Furthermore, event logs may even be lost in the system if there is no mechanism in place to convey them. For instance, it could be challenging to monitor member subscriptions or orders placed if an online business loses event logs including payment or other time-sensitive data.

3.1.2 Metrics: Understanding System Health

Metrics are numerical values that indicate the performance of a system over time. Name, value, label, and date are among the properties that make up the usual format of metrics. This makes searching quicker and simpler and optimises storage, enabling one to preserve them for longer (Crowe, 1993).

Metrics, such as CPU and memory consumption, error rate, network latency, and anything else that provides information about the system's health and performance, are typically the major performance indicators. Response time, peak load, and the quantity of HTTP requests fulfilled are a few KPIs that can be monitored when keeping an eye on a website.

Metrics can be used to determine the severity of an issue, offer insight and visibility, and set off alarms when certain attributes surpass a certain level. Teams can use these indications to gauge the seriousness and performance of the system. For example, one can observe an abrupt increase in system traffic while tracking the number of service requests per second on the website. The rise may be explained by metrics, which could point to malicious activity, improper service configuration, or design flaws. Consequently, aiding in the problem's discovery and evaluation.

Benefits of Metrics

Metrics are primarily used to notify DevOps and other teams of potential mistakes, which can lead to the following advantages:

- **Enhance user experience:** Metrics help gauge how engaged customers are with a business and its offerings. User satisfaction, engagement, and loyalty can be inferred from metrics such as revenue per user (RPU), average order value (AOV), cost per install (CPI), and others.
- **Business execution:** Since metrics are numerical and don't strain the system, they are simple to maintain and query. They

are therefore excellent for dashboards including historical data, particularly for tracking KPIs in real-time and looking for trends in data over time.

- **Less expensive:** Because metrics don't increase in price in response to user volume or other system activity that produces a lot of data, they are more economical than logs.
- **Quick processing even with more storage:** Unlike logs, which rise in response to an increase in application traffic, metrics do not raise disc utilisation, processing complexity, visualisation speed, or operating expenses. Metric traffic can be prevented from increasing at the same rate as user traffic by using client-side aggregation.
- **Alerts:** Compared to executing queries against a distributed system like Elasticsearch and then adding up the results to determine whether an alarm should be sent, metrics are more effective, dependable, and economical when used to deliver alerts.

Metrics can measure system performance over time and offer insightful information, including practical ways to improve system health. Additionally, it offers alarms for real-time system monitoring. For instance, during system outages or overloads, etc. It can also establish new benchmarks for future objectives and identify irregularities.

Limitations of Metrics

Metrics have constraints, just like the other observability pillars. Here are a few of them:

- **System scoped:** It might be challenging to ascertain what is occurring outside of a system due to the system-scoped nature of application logs and analytics. It's also possible to request scoped metrics, although doing so necessitates greater label spread-out and additional metric storage.
- **Performance is slowed by high cardinality:** One user identification is high cardinality data in a system with thousands or millions

of users. High cardinality data slows down its monitoring tools and is more difficult to query. High cardinality monitoring tags might contain thousands of possible combinations. The quantity of processing power and storage required to store all the data may make the system expensive.

- **Multi-format data storage:** Due to the large range of data and analytical techniques needed, using metrics can be difficult. Another issue is how to keep data that has been gathered from different sources and in different formats. However, the work invested in learning how to use the data will be worthwhile.
- **Limited diagnostic capabilities:** Using metrics alone to diagnose an occurrence is difficult. For instance, it is challenging to organize metrics by tags, filter tags, and iterate on the filtering process in a real-world setting where a tag may contain hundreds of data. Furthermore, the cardinality of its data set is greatly increased by the addition of tags.

Metrics vs. Logs

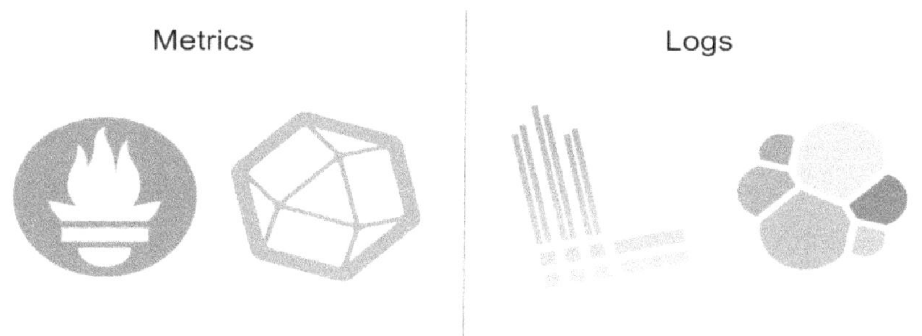

Figure 3.2: Monitoring Tools Metrics and Logs

Source: (Team, 2025)

Despite the fact that both logs and metrics are searchable, timestamped, and offer useful information, their respective benefits are very different.

Metrics are able to track advancement, identify significant events, and predict failures. Because they may apply a simple criterion and are numerical, they are more effective for alerting than logs.

Although debugging is the most popular use for logs, they can also be useful for a variety of other purposes, including tracking program performance and analyzing user behaviour. If you know where to look, logs can assist you figure out what's going on. Logs are great for reactive searches and delving deeper into silos, whereas metrics are excellent for proactive searches and alarms. In other words, logs are ideal for forensic investigation, while metrics are excellent for identifying trends and outliers. For example, logs can provide more information on why a resource ceased functioning.

In the instance of an online business, adding metrics to a dashboard can improve the ability to visualize data and solve problems. They don't need more room or resources as the number of clients rises, in contrast to logs. Although metrics may increase with the number of containers or instances, they are more economical than logs due to their small size.

3.1.3 Tracing: Diagnosing Distributed System Failures

Traces, the final of the three observability pillars, are a means of documenting the activities a user takes while utilizing the application or service. One technique for tracking requests as they pass through distributed systems is called distributed tracing. A distributed trace, for example, tracks a request from the front-end user interface (UI) to the back-end systems and then back to the user. Every user activity, from opening a tab in itsapp to accessing it in itsGUI, is therefore recorded (Shkuro, 2019).

Traces can be used to identify, describe, and prioritize bottlenecks in an application so that they can be optimized. It also aids in the monitoring and debugging of applications that utilize a variety of resources, including disc, network, and mutex. Traces are a crucial component of observability since they give context for the other components. To identify the most helpful metrics or logs associated with the problem you're attempting to resolve, for example, site reliability engineers (SREs) and other ITOps and DevOps teams can examine a trace.

Benefits of Traces

Distributed tracing helps DevOps, ITOps, and SREs accomplish the following goals:

- **Address user complaints:** If a consumer complains about a slow or malfunctioning program, the support staff can examine scattered traces. From the same platform, engineers may then investigate frontend performance issues using an end-to-end distributed tracing tool.
- **Service relationships:** By comprehending cause-and-effect links, distributed traces assist developers in maximizing service performance. For instance, examining a span created by a database request may reveal that a service farther upstream experiences a slowdown when a new database record is added.
- **Monitor user activity:** By using traces, engineers can determine how long it takes for users to complete important tasks like making a payment. They also assist in locating backend bottlenecks and problems that degrade user experience.
- **Boost efficiency and teamwork:** In microservice architectures, many teams may operate the services that comprise a request. Distributed tracing identifies the team in charge of resolving an issue.
- **Uphold service level agreements (SLAs):** In the majority of businesses, SLAs are performance agreements with internal teams or clients. Teams can evaluate SLA compliance with the aid of distributed tracing tools, which compile service performance data.

Limitations of Traces

Traces have their own limitations in terms of performance and implementation, as listed below, even if they can help enhance user experience and identify the application-level underlying cause of issues:

- **human code alteration:** In order to begin tracing requests, certain distributed tracing platforms require human code modification. Although manual instrumentation is often required by the language or framework one wishes to instrument, it consumes engineering time and creates errors. Missing traces can also result from standardizing code instrumentation.
- **Head-based sampling:** At the beginning of each request, traditional tracing platforms randomly select samples of traces. Sadly, this approach only leaves partial evidence. Furthermore, important business traces like high-value transactions or enterprise client requests may be overlooked by this head-based sampling.
- **Only backend coverage:** A trace ID is only generated for a request once it reaches the first backend service if an end-to-end distributed tracing platform is not used. User sessions on the frontend are not visible. Determining the root cause of a request and whether a frontend or backend team should address it becomes more difficult as a result.
- **Various programming languages:** If a system employs several programming languages or frameworks, tracing may be challenging. Since all functions transmit traces, developers might have to find tracing solutions for every language or framework in a system. Retrofitting heterogeneous systems with tracing is more difficult.

Traces vs. Logs

In order to centrally monitor error reports and associated data, logging is the process of collecting and storing data an application generates when operating normally. The program activity is the main focus. System administrators can therefore use comprehensive logs to address application issues.

From endpoint to endpoint, distributed tracing contextually tracks a single transaction. It aims to identify the precise location of a mistake. By emphasizing service communication, it assists teams in recognizing service interdependencies. As a result, troubleshooting will take more time while identifying the problem's origin takes less time. It also assists businesses in identifying and fixing app performance problems before users become aware of them.

Although they don't show source code faults, traces are a richer source of background information than logs. Information visualization expedites the process of identifying and resolving issues by automatically creating traces. Although they are less flexible and adaptable than logs, traces delve deeper into distributed systems.

Table 3.1: Logs vs Traces

Feature	Logs	Traces
Granularity	Detailed records of events	High-level overview of request paths and detailed records of activities (in the form of spans)
Use Case	Compliance / auditing	Performance analysis, service interactions
Data Volume	Typically larger, continuous streams of multiple logs	More structured, and therefore, denser: many attributes in a span
Dealing with Data Volume	Look for cheaper storage to minimize costs	Sample traces, dropping redundant ones
Context	Limited context of system behavior	Rich context of interactions
Level of Effort	Less work to implement	More work to implement

Source: (Team, 2025)

Here are some general guidelines for when to choose each:

- **Logging:** Logging can be used to capture events and messages in a program, including status updates, error messages, and debugging data. Additionally, logs can be used for compliance and auditing.
- **Traces:** Select traces if you need to comprehend how requests go through the system and how each component functions. Trace is especially helpful for locating bottlenecks and diagnosing performance problems.
- **Metrics:** When one has to monitor the system's performance over time and spot trends and patterns, metrics are the best tool to utilize. Metrics can be used to assess progress, define performance goals, and optimize an application based on data.

3.2 Setting Up Effective Monitoring Systems

Monitoring within a company or supply chain is typically categorized into two main types: sporadic monitoring and ongoing monitoring. Sporadic monitoring is conducted at specific points in time to assess compliance with defined standards and requirements. A common example of this is an audit, which evaluates whether certain conditions are being met on the day of inspection. This may include checking if employees are wearing appropriate protective clothing, ensuring adherence to fire safety protocols, and verifying that wage payments are being made regularly and correctly. Sporadic monitoring offers a snapshot of compliance but does not capture changes or trends that may develop over time (Blanchard, 2021).

In contrast, ongoing monitoring involves the continuous observation and assessment of a supplier or a company's internal operations over an extended period. This approach focuses on identifying and addressing specific risks while tracking key performance indicators (KPIs) to drive long-term improvements. Unlike sporadic monitoring, which offers a momentary look at compliance, ongoing monitoring builds on the results of audits, self-assessments, and other tools to form a more holistic view. It integrates various sources of information, such as channels for complaints, advice from quality control staff, and comments from staff training initiatives, to offer a more comprehensive grasp of the working environment.

In addition to assessing compliance, a well-designed continuous monitoring system helps employees and suppliers develop their capabilities. This entails giving stakeholders focused assistance and chances for growth so they can progressively enhance their procedures. As staff members and suppliers gain the ability to uphold high standards on their own, such systems can eventually lessen the need for continual supervision by encouraging a culture of continuous improvement. This proactive strategy guarantees that any problems are resolved before they become significant ones and that the company gradually advances towards greater performance and accountability levels.

Moreover, the information gathered through both sporadic and ongoing monitoring can be invaluable in evaluating the effectiveness of a company's management systems. For instance, analyzing long-term data on workplace accidents can help determine whether health and safety training is successfully reducing incidents. Similarly, trends in excessive overtime can highlight the unintended consequences of procurement and sourcing strategies, prompting necessary adjustments to protect worker well-being. Through this feedback loop, monitoring not only ensures compliance but also becomes a strategic tool for continuous organizational development and improvement.

3.2.1 Key Performance Indicators (KPIs) to Monitor

Specific indicators of success (KPIs) are part of a strong monitoring process that assist one determine whether the company, its partners, and its producers are meeting their objectives. The input, output, and result KPIs are distinguished. The resources used for a particular production step or goal are measured by input KPIs, and the outcomes of a process are measured by output KPIs. The goal of progressive monitoring is to track real progress, or result KPIs (Wiley, 2020).

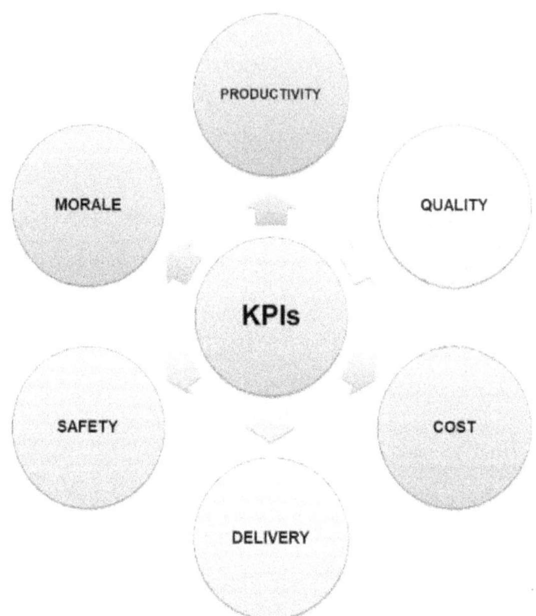

Figure 3.3: Key Performance Indicators (KPIs) and Their Impact Areas.

Source: (TQP, 2022)

In contrast to output KPIs, outcome KPIs do not take into account whether or not a measure has been put into place, such as introducing a complaint channel or offering training on handling chemicals. Instead, they concentrate on the results, such as whether the training has improved chemical handling or whether staff members believe that the company's complaint procedures are a valid and easily accessible tool. Outcome KPIs frequently need the collection of indirect indicators and/or qualitative (descriptive) data because they cannot be measured directly.

Since they serve as the foundation for all ensuing data collection, processing, and reporting, indicators are an essential component of a monitoring system. Monitoring and evaluation (M&E) efforts are unable to compare actual progress with predicted and agreed-upon progress if they lack a clear set of indicators. At every stage of the results chain, they are necessary. “Key performance indicators” are metrics that show the level of outputs, results, and higher-level objectives.

Utilization of resources is measured by input indicators. They monitor if the necessary resources are available to carry out the intervention. X budget allocation and expenditure are examples of common input indicators. X number of local facilitators under contract X number of local organizations that contribute in-kind X number of program workers by level X amount and share of matching funds have been raised.

3.2.2 Choosing the Right Monitoring Tools

For each indicator used in monitoring and evaluation, it is crucial to determine the most appropriate method of data collection. In general, this entails selecting between the two primary categories of data: qualitative and quantitative. Numbers that can be measured, categorized, ranked, or statistically evaluated are known as quantitative data. These statistics are particularly helpful for recording economic situations, demographic information, and other quantifiable markers. Standardization and impartiality are given top priority in quantitative approaches, which enable uniform data collecting across various populations or time periods. Tools such as surveys, structured questionnaires, tests, and national or organizational censuses are common in this approach. These instruments often rely on closed-ended questions, making them effective for generating large datasets that can be compared and analyzed for trends or patterns (Konrad Hugo Jarausch, 2021).

In contrast, qualitative data are non-numerical and focus on exploring the meanings, motivations, attitudes, and experiences behind people’s actions and perceptions. These data help explain the “why” and “how” behind behaviours and outcomes, providing rich, in-depth insights into stakeholder perspectives. Qualitative methods include unstructured or semi-structured interviews, focus group discussions, case studies, ethnographies, and direct observation. These approaches are particularly useful in contexts where cultural, social, or emotional factors play a significant role in influencing outcomes. While qualitative methods often yield more nuanced information, they can also be more subjective and may require careful interpretation to avoid bias. However, they tend to be faster and more cost-effective to implement, especially in settings where large-scale quantitative data collection is not feasible.

An effective evaluation strategy often involves integrating both quantitative and qualitative methods, a practice known as the "mixed-methods" approach. By combining the strengths of both types of data, organizations can gain a more complete understanding of a programme's implementation and its impact. For example, quantitative data might reveal a drop in school attendance, while qualitative interviews with students and teachers might uncover underlying causes such as bullying or poor infrastructure. This integrated approach enhances the validity of findings and supports more informed decision-making by providing both statistical evidence and contextual understanding.

Ultimately, the choice between quantitative, qualitative, or mixed methods depends on the specific objectives of the monitoring activity, the nature of the indicators, available resources, and the context in which data collection will occur. A thoughtful selection of data collection methods ensures that the information gathered is not only accurate and reliable but also meaningful and actionable, thereby strengthening the overall effectiveness of monitoring and evaluation efforts.

2.2.3 Implementing Real-Time Alerts and Dashboards

In the modern data driven environment real time dashboards have been emerging as indispensable topline, instant knowledges about operation, system and user interactions. Real time dashboards are as opposed to the traditional business intelligence (BI) dashboards that maintains the data and refreshes it on the scheduled or batch basis. In domains where real time decision making is crucial, these systems have a dynamic, up to the second visualization of key metrics that are invaluable. They are an architecture that allow multiple users to access concurrently and guarantee that all the viewers see the most recent information that is available (Eckerson, 2012).

When it comes to a real time dashboard, the data sources are at its very core. The continuous feeds of information from different systems, devices and services are these sources. For example, in an industrial setting, the sensor values may be pulled in real time from machinery in the form of real time dashboards. They could track website traffic, sales transactions, social media interactions as they happen in digital application. This allows these dashboards to be real time, because they can integrate into a very wide array of data sources each of which provides high frequency updates.

Further, as a critical intermediary component, the data processing engine takes raw input and transforms it into something useful, in this case, the output provided to the user. The incoming data streams are aggregated, filtered and formatted by this engine, to be consistent and preparation for visualization. It must be very efficient and be able to handle large volume of data with small latency. The processing engine also runs business logic or real time analytics such as anomaly detection or trend forecasting on top of the dashboard enhancing its value greatly in most systems.

Frontend of the dashboard, the visualization layer is the part that the users see and interpret the data through. The interactive graphs, charts, heat maps, and so on, which are part of this layer, automatically update as the incoming data changes. On top of presenting the information in a usable fashion, a good visualization layer also allows users to pick up patterns and correlations and also detect anomalies at a glance. What makes these visualizations real time is it allows organizations to detect whether or not there are issues, and to seize on opportunities, as they come about.

Interactive controls embedded inside of the dashboard interface are equally important. These controls allow one to filter data, drill down to more granular views on the data, compare historical trends and alerted on specific events. A good example is: a user may isolate a given set of data from a specific area, monitoring only critical metrics or receive a notice when a KPI reaches a given threshold. In an effort to make data exploration experience more personalized, responsive and even more important — fast and with greater confidence — these interactive features were developed.

Finally, real time dashboards are more that modern ting visualization, real time dashboards are powerful, respond tools integrated live data streams, real time processing aficionados, and user critical controls to give without delay, swiftness, insights. They continue to become more and more important to situational aware, operational efficiency and strategic agility in all industries.

3.3 Incident Detection and Response Strategies

Incident detection and response are critical components of an organization's cybersecurity framework, encompassing the processes and activities involved in identifying, analyzing, and reacting to potential security threats or incidents within the IT infrastructure. Effective incident detection requires the continuous and proactive monitoring of networks, systems, and applications to identify anomalies, suspicious behaviours, or known indicators of compromise. This vigilance ensures that threats are identified at the earliest possible stage, minimizing the potential for damage and disruption. Detection mechanisms can include intrusion detection systems (IDS), security information and event management (SIEM) tools, endpoint detection and response (EDR), and behavioural analytics (Thompson, 2018).

Once an incident is detected, a timely and coordinated response is essential to contain and mitigate its impact. Incident response strategies typically involve a structured approach, including preparation, identification, containment, eradication, recovery, and lessons learned. This cycle not only helps in resolving incidents efficiently but also contributes to the continual improvement of security posture through post-incident analysis and refinement of response protocols. An effective incident response plan should clearly define roles and responsibilities, communication protocols, escalation paths, and coordination with external stakeholders such as regulators, law enforcement, or cybersecurity authorities, depending on the severity and nature of the incident.

Under the NIS 2 Directive, several key requirements align closely with incident detection and response. These include obligations for essential and important entities to implement risk management measures that encompass incident handling capabilities. Sections within NIS 2 emphasize the need for timely incident reporting, coordinated disclosure of vulnerabilities, and maintaining adequate logging and monitoring capabilities to facilitate early detection and analysis. Additionally, the directive mandates the establishment of a robust cybersecurity risk management culture, which includes equipping organizations with the tools, processes, and personnel necessary to detect and respond to incidents effectively.

Furthermore, NIS 2 encourages the development of cross-border cooperation frameworks for incident response, particularly for large-scale or transnational threats. Organizations are expected to participate in national or sector-specific Computer Security Incident Response Teams (CSIRTs), which provide expertise, guidance, and coordination support during major security events. By aligning incident detection and response strategies with the NIS 2 requirements, organizations not only ensure regulatory compliance but also strengthen their resilience against evolving cyber threats. For additional guidance on continuous monitoring or vulnerability management practices, refer to the relevant sections of this guide, which provide deeper insights into proactive security measures that complement incident detection and response efforts.

3.3.1 Automating Incident Detection with AI

Incident response with AI implies the use of Artificial Intelligence (AI) and the Machine Learning (ML) technologies strategically to analyses and detect security incidents in real-time. Consequently, this approach has completely redefined traditional cybersecurity process by automating some of the processes and reducing man hours greatly. The growing cyber threat makes it more vital than ever that incident response systems can operate with greater speed, accuracy and consistency, and that has to do with AI at its core (Digital, 2024).

This is the beginning of integrating AI into the incident response process and bringing the security postures from reactive to proactive. Large number of security data can be processed with sophisticated algorithms which can detect patterns and anomalies that can signal malicious activity. It helps organizations to foresee and nullify threats before they become horrendous breaches. This means enterprises can effectively protect their business operations from any attack and improve the overall strength of their defense network.

Several core functions make AI powered systems transform the incident response service. Automated threat detection first uses the power of artificial intelligence powered, behavioral analytics to stop the flow of automatic suspicious activities across digital environments. They can also detect subtle indicators of compromise something that might not otherwise be noticed by human analysts. Second, intelligent analysis of security alerts helps teams prioritize incidents as they can become only the most severe and high impact threats. Third, threat containment is accomplished automatically by isolating affected systems to stop the spread of cyberattacks across networks. To finish, AI enabled execution of pre-defined incident response playbooks enables security teams to quickly and effectively respond to a breach.

There are multiple and compelling benefits of incident response using AI. The most significant advantage is the fact that threat detection is much faster and significantly decreases the time it takes to determine and assess possible attacks. It is this speed that's important to reduce the damage caused by fast moving threats like ransomware. This also reduces the number of false positives. The use of AI improves detection accuracy and thus allow the security personnel to be freed up to handle the real threats and not alerts. Furthermore, automation reduces the manual work for cybersecurity teams, and facilitates how their resources are allotted.

Another important benefit is scalability since AI systems are capable of processing and analyzing large datasets in real-time, which is the key characteristic for large enterprises of complex and constantly changing IT architecture. AI therefore leads to a proactive security stance by supporting predictive analytics that predict exploitable vulnerability and exploit vectors before the attacks occur. This foresight capability enables AI not just to enable quick-response to incidents, but also integral part of the preventive cybersecurity strategies.

However, it is worth mentioning that AI supported incident response has revolutionized incident response for organizations controlling digital assets. It is given intelligence, speed and automation in the fabric of security operations so that it becomes more resilient, more efficient and more proactive in responding to the ever-evolving threat landscape.

3.3.2 Categorizing Incidents: Critical vs. Non-Critical

The category of incidents is key to effective incident management; more specifically, critical and non-critical. It is this classification which helps the organization to stream line its response strategies to attend to the critical incident immediately and to handle less critical incidents according to standard procedures. This clearly defines severity of each incident so that it is not confused, does not over allocate resources and has a structured approach for incident resolution (Vincent Faggiano, John McNall, Vincent/McNall Faggiano (Joh), 2011).

Such critical incidents include those that threaten business continuity, integrity of data or user experience. In these cases, system outages, major security breaches or application failures which directly impact a lot of users or key business processes are involved. The immediate aftermath from not responding to such incidents can be equally as destructive financially and with regards to regulatory noncompliance and reputational damage. Thus, the critical incidents need to be responded to immediately and multiplicatively, which is accomplished through escalation protocols, cross functional working and pre-defined emergency workflows to avoid downtime and reduce risk.

On the other hand, non-critical incidents are issues that do not have immediate impact on the core operations of an organization or its users. Examples of these would include minor bugs, low priority service requests or performance degradations which have acceptable workarounds. Most of these incidents are handled during the regular business hours and do not require the disruption of normal workflows or engagement of emergency resources. However, low urgency non critical incidents should also be logged, monitored and resolved systematically to prevent them from becoming critical incidents over time.

Table 3.2: Critical vs non critical incidents.

Aspect	Critical Incidents	Non-Critical Incidents
Definition	Incidents that threaten business continuity, data integrity, or user experience.	Incidents with no immediate impact on core operations or users.
Examples	System outages, major security breaches, application failures.	Minor bugs, low-priority service requests, performance issues with workarounds.
Impact	High impact; can cause financial loss, reputational damage, and regulatory non-compliance.	Low impact; unlikely to disrupt operations if addressed during standard hours.

Aspect	Critical Incidents	Non-Critical Incidents
Response Strategy	Requires immediate, multi-layered response through escalation protocols, cross-functional teams, and emergency workflows.	Handled during regular business hours without disrupting normal workflows or involving emergency resources.
Resource Allocation	Demands prompt attention from specialized resources (e.g., security analysts, IT teams).	Managed with standard resources; less urgent but still requires monitoring.
Automation	May trigger predefined emergency playbooks and immediate escalation paths in automated systems.	Can be queued and routed based on priority via automated incident management systems.
Operational Outcome	Enhances resilience by reducing downtime and mitigating major risks.	Prevents issues from escalating; supports continuous improvement.
Customer Experience	Fast resolution boosts transparency and customer trust.	Proper handling ensures reliability and promotes service consistency.
Organizational Benefit	Maintains business continuity and builds a reliable reputation.	Encourages a culture of accountability and sustained service quality.

Source: (self-generated)

An accurate classification also makes resource optimization possible. Finding out which incidents matter can help organizations to use their fixed infrastructure resource for incident response: security analysts, IT personnel or support engineer. With this, teams are not overwhelmed with low priority tasks but important incidents receive adequate attention too. Furthermore, automated incident management system can be set up to handle incidents automatically and route, or escalate an incident based on its severity.

Therefore, a well-organized incident categorization framework helps in increasing operational resilience and customer satisfaction. Customers do not experience much disruption when critical issues are solved quickly and in a transparent manner, and they maintain trust in the organization's reliability. It also helps to systematically manage non critical incidents to improve continuous service and promote an accountability culture. The companies who can narrow the criteria to name the three most critical trades at the start of a day, and clearly differentiate those three from all the others, will find themselves in a position to have real opportunity.

3.4 Automating Incident Response and On-Call Best practices

Today, organizations have a continuous expectation from the availability of their systems and applications to secure customer demands and develop user experience. Financial losses, customer dissatisfaction, damage to the brand's reputation can occur due to any unplanned downtime or disruption. Incident response has become a strategic imperative and automating it is the way to handle these challenges. Integration of Automation into Incident Management allows organizations to increase detection, analysis and resolution of incidents, thereby minimizing the loss of revenue due to instances of outages. When combined with appropriate on call best practices, application of this automation provides business with resilient, proactive, and effective mechanism to respond to incidents while supporting employee wellbeing and operations (Sheffi, 2016).

Automated incident response is using predefined rules, scripts and intelligent algorithms to respond or react incidentally either in or near real time. The heart of this approach is that it is possible to constantly monitor systems and environments for abnormalities using advanced telemetry, log analysis and performance metrics. These monitoring tools are always on and willing to attempt to detect these unusual patterns, generate alerts and workflows without needing human intervention. For example, if a critical application gets crashed or latency rises to more than a given threshold, the affected application's teams get automatically notified and the related services start to be restarted, isolated components, or infrastructure is scaled dynamically to prevent performance. Manual methods alone can not achieve this kind of agility and significantly increase system reliability.

Several components have to be integrated to make automated response effective. Event correlation is key to collecting data from disparate sources to understand the whole context of an incident. Organizational knowledge encoded in playbooks (or runbooks) means that even complex incidents are handled in the same consistent and efficient manner. Moreover, artificial intelligence (AI) and machine learning (ML) models add in automation by increasing automation through continual learning of past incidents, increasing the alert accuracy, and predicting future failures. Also, these automation systems are integrated into tools such as Slack, Microsoft Teams, or PagerDuty using the ChatOps principles to accelerate collaboration, making it possible for teams to coordinate response tasks from one place.

While automation is beneficial in many ways, the incidents that are new, ambiguous, or of high business impact will always require human judgment for incident management. They are in place as the safety net in case automated systems fall short or a confirmation is needed before a sensitive action is carried out. For the effectiveness, organizations must be implementing comprehensive on call practices. This includes that there are clear escalation paths, with who is responsible and when they need to respond to different types of incidents. Equitable rotation schedules mean that workload is distributed evenly across teams without burning out the team responsible for the same for weeks or even months. Alert hygiene is another big component that involves tweaking alert radians, lowering noise and lowering false flags so that engineers can concentrate on actual troubles rather than getting drowned out. Providing runbooks, diagnostics and service dependencies easily accessible to on call engineers increases their ability to respond quickly and confidently.

When these two are in sync, there is a high functioning, responsive incident management framework when automation and on call practices are aligned. Common and repetitive incidents are handled by automated systems, which cuts out mean time to detection in leaps and bounds and relieves engineers to focus on their high-level critical tasks that demand their expertise. At the same time, on call protocols assure that human responders are equipped with the right tools and information and are empowered and supported. The dual approach to this increases operational efficiency, decreases downtime and amplifies service level objectives (SLOs). In addition, automation logs, on call metrics can be extracted to find bottlenecks, improve workflows and prevent such incidents in the future. Organizations create a culture of continuous improvement that makes systems resilient and becomes competitive by delivering digital services that are of high quality, high availability, and continuity.

Figure 3.4: Incident response tools

Source: (Squadcast, 2023)

The image illustrates key components of incident response tools, which are vital for managing and resolving cybersecurity incidents. These tools help organizations detect threats, analyse vulnerabilities, respond to attacks, and recover from security breaches. The components are:

- **Alerts**: Notifies security teams of potential incidents, enabling prompt action.
- **Automate**: Streamlines response processes, reducing manual effort and response times.
- **Analyze & Report**: Helps in understanding the nature and scope of incidents and provides documentation for future prevention. Security Information and Event Management (SIEM) systems are used to aggregate and correlate security data from multiple sources, which helps teams identify and analyze potential incidents more effectively.
- **Integrate**: Ensures compatibility with other security systems for a coordinated response.
- **Collaborate**: Facilitates communication and teamwork among incident response team members. Incident management tools often include collaboration features for real-time communication among team members.
- **Training & Support**: Equips personnel with the necessary skills to handle incidents effectively. A Computer Security Incident Response Team (CSIRT) is crucial for providing staff training and maintaining security awareness.
- **Scale**: Allows the incident response system to adapt to the size and complexity of the incident.
- **Customize**: Enables tailoring the tools to specific organizational needs.

Incident response must be timely, automated, analyzed, integrated and collaborative, which is supported by training and customization for organizational needs. SIEM and CSIRT are tools to enhance scalability and preparedness to various security threats.

3.4.1 Reducing Manual Intervention with Automation

Automation for its part reduces manual intervention and may entail picking repeating task, improve workflows and using new tech in data entry, approvs and communication. Through these processes streamlined, organizations can greatly improve operational efficiency, lower errors and allow staff to invest more time to concern, other strategic, additive value tested processes. The more natural one takes to automation, the faster they can automate routine activities, and the more consistent as well as accurate those routine activities will be, especially in systems and activities with multiple systems and teams (Christie, 2012).

The positive impact that can be achieved on the whole supply chain is one of the key benefits that come from minimizing the manual intervention. This can become a cascading effect that will disrupt the flow of goods and services for any delay in repairs, approvals or communication. Not only do these delays break down internal teams, but they can incur frustration for the customer and reduce customer satisfaction and experience. Manual assignment can be replaced with digital solutions, which means tasks can be sent automatically to each operative. This helps everyone on the chain to have access to the necessary information, monitor progress in real time and coordinate more efficiently. Transparency is also increased through automated systems, which makes it possible to better plan and schedule, and provides better delivery windows that ultimately mean better service levels to the end customer.

The elimination of human error is another big advantage in reducing manual intervention. There is simply the manual process which is extremely prone for mistakes like entering wrong figures, missing updates or wrong communication of key information. For instance, little things such as a mismatch in pricing or data entry can result into large loss and delays in costing. These elements need to be automated to ensure that data syncing happens in real time between systems and minimizes the risk of errors and increase data integrity. Accurate and up-to-date information allows us to bring more confidence when making informed decisions for the good of entire organization as the stakeholders can rely on it.

Automation ends up being a force for operational efficiency as well as organizational resilience. Today, the ability to react quickly and soundly is a winning playing card. Such businesses can make their processes more agile and scalable processes that adapt to changing demands due to being less dependent on manual tasks. Automation in industries is no longer an option, it is an essential part of business to maintain accuracy, accountability, provide excellent service, and continue digital transformation.

3.4.2 Managing On-Call Rotations Effectively

An on-call rotation is a structured schedule where the members (mostly SREs) are assigned for availability outside of normal working hours to respond to system incidents in order to maintain service stability. These are on call engineers who are first line of defense during system issues and were responsible for resolving alerts, diagnosing problems, and escalating the issues as needed. On Call responsibilities are usually about 25% of an SRE's time (i.e. 1 week per month). But in managing an on call shift the assignment is not the whole equation, it should still have a thoughtful schedule, balanced workloads, optimized alert system, and have a high reliance and accountability culture (Niall Richard Murphy, Chris Jones, 2016a).

The key elements of successful on-call management begin with the design of those on-call schedules that balance on the one hand system coverage and on the other hand the wellbeing of the engineers. Shift composition must be defined; engineers must know clearly what their roles involve, frequently monitoring systems, fixing problems and performing incident resolutions. The runbooks need to be updated continuously to include troubleshooting steps, commands, and procedural information so that on-call engineers have always access to up to date and relevant guidance during incidents. Handoff procedures are just as important—shifts should not have to stop what they are doing to catch up with what has gone on, and details should be provided on current system status, unresolved issues and completed tasks. In addition, post mortem meetings should be conducted regularly to analyze incidents, find out root causes and take corrective actions to avoid recurrence.

One of the most important aspects of on call management is improving its alerting procedures to reduce noise and reduce alert fatigue. Pager load optimization is the art of reducing the resolution required for an alert from something very uncommon that deserves full blown attention to something common and unlikely to be much of a problem. Prometheus, Datadog, or PagerDuty type of monitoring tools are great for not only having read outs of the system health but rather tracking in real time when alerts should go out to the right people. Also, the escalation plans should be documented well and able to be executed easily, identifying the specific routes for transferring responsibility to higher level support when incidents become complex or high impact. This eliminates the need for critical issues to wait in line, to be addressed by the wrong experts at a later date.

Supporting on call operations requires training and documentation. There are also new SREs that have to be completely on boarded, in the form of being technically onboarded, through tools, incident management protocols, techniques that troubleshoot. All team members are trained on an ongoing basis and are made to simulate high pressure situations. Shadowing opportunities with experienced engineers can encourage the on-call process to become familiar to newer team members and equitably transfer knowledge within the team.

To implement best practices, organizations should take on a few foundational strategies. To achieve effective handoffs, shift reports must be made known to one another on platforms like Slack, or perhaps on dedicated incident management tools. A culture of learning is fostered by weekly post mortem analysis which looks at recent events and makes the process better. Keeping runbooks up to date enables rapid resolution of issues even in the face of stress. So for this, finally, one can spend its money in such scalable and user-friendly tools like Squad cast or

Opsgenie, to route the alert through, to manage escalation, and to keep it call system well-coordinated. All of which help in providing operational resilience, decrease operational downtime, and improve the reliability of the production system.

3.4.3 Burnout Prevention for On-Call Engineers

Burnout is a condition that occurs from prolonged and chronic stress which is usually the result of demanding work conditions. Burnout is a very debilitating problem for many professionals, especially in technical and operational positions, bringing down productivity, disengagement, even long-term health impact. For site reliability engineering and for incident response, the responsibilities draw engleers to this risky state of burnout naturally because of the work they are doing: they need to be alert, responsive and calm under pressure, sometimes with no opportunity to rest and recharge (Nancy McCormack, 2013).

Burnout is a common and serious problem for on call engineers in particular. Unexpected alerts at midnight are a fact of life for them, forcing them to wake abruptly and quickly enter complex problem-solving situations. The need to react quickly and make decisions is crucial in these high-pressure incidents, and many have high potential impact to the business in a short window of time. Add to that the fact engineers are likely to never have downtime, and there is the added mental and physical strain of having to deal with these emergencies hand over hand.

On top of the technical demands, on call work can really interfere with its personal time and wellbeing. They can be interrupted by urgent pages to their engineers, leaving them feeling imbalanced and fatigued all of the time. If not supported with sustainable scheduling, recovery mechanisms, disengagement or lowered morale are very likely. Organizations must take on humane work practice and do things like rotate shifts fairly and manage alert systems to reduce noise so engineers have a decent work life balance when not on call.

How to Prevent Burnout in On-Call Engineers

1. Limit How Often Engineers Are on Call

Rotate on-call duties fairly so that no one is overwhelmed. Give people breaks between shifts so they can rest and recharge.

2. Make Sure Alerts Are Worth It

Reduce unnecessary alerts (also called "alert noise"). Only wake engineers up for real emergencies. This helps them sleep better and feel less stressed.

3. Support Recovery After Incidents

If someone handles a tough issue late at night, give them some time off the next day to rest. Don't expect them to be fully productive immediately after a 3 AM outage.

4. Provide Mental Health Resources

Give access to support like counselling, therapy, or wellness programs. Let people know it's okay to ask for help.

5. Encourage a Healthy Culture

Make it normal to speak up if someone is feeling burned out or overwhelmed. Teams should support each other, not blame each other when things go wrong.

6. Use Automation and Tools

Automate common tasks or fixes so engineers don't have to manually solve the same problem over and over. This reduces the workload and stress.

7. Recognize and Appreciate On-Call Work

Let engineers know their efforts are seen and valued. Celebrate good incident handling, and include on-call duties in performance reviews.

3.5 Postmortems: Learning from Failures

A post-mortem, also called a lesson learned meeting or incident review, is an all too structured and reflective process that takes place after a project, incident or operational failure. However, its main purpose is to study what happened, the reasons for failure, and what can be done to avoid it happening again. Instead of seeking to blame, the post-mortem process focuses on collective learning and systemic improvement and, therefore, it is a necessary process for teams working on projects of project management, engineering or site reliability operations (Weinberg, 2001).

A post-mortem has a broad and flexible scope. Post-mortems, so commonly associated with outage or technical incident analysis, can also be performed around project delays, misaligned processes, etc., or even unexpected successes. The essence is the same no matter what the case is: learning from the past to improve the future. This practice fosters a culture of continuous improvement, accountability, and transparency across the organization.

The blame free environment is considered to be one of the most important elements of a successful postmortem. Psychological safety promotes a discussion about mistakes with less fear of being reprimanded and teams can feel more psychologically safe to share their perspectives with each other. This is to get at the full scope of the problem and to understand the human and systemic factors involved.

An important part of the post-mortem process is to root cause identify. The teams should look beyond mere surface level symptoms or immediate failures to seek the root causes. The employing of techniques such as the "Five Whys" or fault tree analysis are able to trace back the chain of events, whereby it comes to reveal weaknesses in systems, systems of communication and even the workflows themselves.

The final stage of the postmortem process should be to produce a concrete action plan. Other lessons identified must be translated into namely defined, action steps itcan take to reduce risks and increase future performance. They should be well documented and shared across relevant teams and these outcomes as well as details of the analysis should be documented. Documentation helps to retain valuable insights through personnel change or change in the project itself, eventually leading to the establishment of a resilient and proactive organizational culture.

3.5.1 Conducting Blameless Postmortems

A blameless postmortem is a foundational practice in a world class technical team in particular such within Site Reliability Engineering (SRE) and DevOps cultures. A blameless postmortem aims to create a culture where, instead of people being afraid to discuss incidents, failures, or outages (and thus, outages persist), they can talk about incidents, failures, or outages without fear of firing or finger pointing. Transparency, responsibility and iterations of improvement, not defensiveness, or concealment (Ukis, 2022).

A blameless postmortem shift from the error of the individual to the systemic understanding. The root cause for human error is not human error, but rather process gaps, unclear documentation or flawed assumptions. This mindset brings out the root causes of the incident, so organizations do not punish people but rather tackle the actual reasons. Without blame, teams are safer reporting issues and also more likely to contribute truthfully to post incident analysis.

To do a successful blameless postmortem, first define a timeline of the event including what happened, in what order, what systems did, and what decisions were made. Involve all the relevant stakeholders to get several perspectives and get a better idea of what happened. Reconstruct the incident and what areas could improve without judgment using the tools that enable collaboration.

The next step is analysis of root cause. The alternate is to ask: Who made the mistake? This brings us to ask "Why did the system bring this mistake?" And often, this insight is that the team is not trained, there are no runbooks, alert fatigue, or misconfiguration of the systems. Apply this gold of insight and create a set of actionable improvements; raise the documentation, adjust alert thresholds, or creates better testing practices.

The findings and action items should finally be documented thoroughly and should be disseminated to the team or organization. To learn more than to blame, others can now learn from the incident with this transparency. Blameless postmortems not only helps to solely prevent such failures occurred again but also fosters trust, deepened collaboration, and adds to the engineering culture to be more resilient.

Benefits of Post-mortems:

Figure 3.5: Benefits of a blameless post-mortem.

Source: (Bigelow, 2021)

- **Improved team morale and dynamics:**

By fostering a safe environment for open communication and learning from mistakes, blameless postmortems can strengthen team cohesion and morale.

- **Faster, more effective corrections:**

Analyzing incidents without the pressure of blame allows teams to identify root causes more efficiently and implement targeted solutions.

- **Enhanced team collaboration:**

Postmortems encourage collaboration by bringing together different perspectives and expertise to understand and address issues collectively.

- **Better product and service quality:**

By learning from past incidents and implementing improvements, organizations can enhance the reliability and quality of their products and services.

Blameless postmortems are considered a key element of a healthy engineering culture, promoting continuous improvement and learning from both successes and failures.

3.5.2 Documenting Lessons Learned for Future Prevention

Any high performing organization will need to document lessons learned after an incident or failure. It is not just about recording the things that went wrong, but about setting up a good feedback loop that will help to prevent future same kind of issue. However, when properly managed, this document turns into an ever-evolving knowledge bank, that enables continuous improvement, increases operational resilience and provides insight on how decisions might be taken in other circumstances (Karl E. Weick, 2015).

It is important to take the first step to capture all critical information in a structured and comprehensive way after an incident. Among them is a timeline of events, a description of the systems or components involved, initial signs of failure, investigative steps taken, and how the problem was eventually resolved. The crux of a SAE occurs in context, thus technical data (e.g., logs, system alerts) should join with human context (e.g., decisions of agents making paths of response, communication that shaped viewing of the response path). All of these elements combined make a complete picture which is invaluable when going back to the incident or training new team members.

Good documentation includes not only telling what happened but going above and beyond to do the root cause analysis. By becoming deeper than the symptoms, teams can not only identify them but uncover deeper system or process flaws. One can learn actionable insights from there. This might have been resolved by changing testing protocols, adjusting alert thresholds, or bettering the team coordination when faced with critical failures. The specific, measurable, and assigned to a responsible party for follow through should be every recommendation. To take accountability, the first step is inaction without ownership, often times it leads inaction, accountability is a core part of the process.

The lessons learned are only valuable if they can be accessed. Thus, it is important for organizations to make sure that all the postmortem documentation is stored in a centralized place that is easily searchable such as a shared wiki or a documentation portal or a dedicated incident management system. It should be categorized in such a manner that it can be retrieved easily later, e.g. by service, by incident type, by severity. This allows teams to view past learnings before making system changes or new features within a team, diminishing the opportunity to make the same mistakes again.

In the end, this is a cultural shift where a continuous learning and shared ownership of reliability and reliability is being practiced. The idea of failure as an endpoint is reinforced by documenting lessons learned. Teams that are invited to reflect without the fear of blame and learn from every incident build a culture of improvement, transparency and resilience. Such a mindset not only helps the team perform better in the present but also prepares them to tackle challenges in the present and in the future with confidence.

3.5.3 Building a Culture of Continuous Improvement

However, it is very important to create a culture of continuous improvement in any organization, and this is even more critical in fast environments like software engineering, operations or reliability engineering. This is a culture at its core serving teams to evaluate their performance on an ongoing basis, learn from experience (especially mistakes) and iterate their processes and systems. Continuous improvement is not one time and that commitment shall continue no matter how perfect its current state of excellence, learning and innovation is (Ruggles & Harrington, 2018).

For its part, much of this culture is based on embracing a growth mindset. A common mistake it make is assuming that resilience is associated with enduring extenuating ordeals without game playing, but those conditions actually make people and firms less resilient and more prone to burnout and break down. With a growth mindset, all these are valued and therefore critical for improvement. An organization must give its team members the safe spaces to openly talk about what went wrong without reprisal.

Structured reflection is also going to be part of a successful continuous improvement framework. They should do retrospectives, incident reviews or performance evaluations at regular intervals to uncover how improvements can be made. Blameless post-mortems that look at the root causes and the system wide failures rather than attributing someone's personal responsibility to this can make this easier. Indeed, these sessions are most useful when they are documented and followed by actionable steps to resolve identified issues.

Practicing culture depends on leaders. Leaders who take an active role in supporting improvement, contribute resources for experimentation and acknowledging its efforts to change contribute to the reinforcing of the importance of continuous learning. Also, metrics should be set up clearly and data analytics can be used to allow teams to make informed decisions, measure progress and refine their strategies.

Finally, a culture of continuous improvement is more than a process optimization – it's about equipping people to be accountable for the work, identify better solutions and consistently improving the standard of work. Such mindset within organizations not only makes them more efficient and reliable but also culminates in a building of stronger, more engaged minds and healthy teams that are confident enough of serving the challenges of tomorrow.

Multiple Choice Questions (MCQs)

1. **Which of the following is not one of the three pillars of observability?**
   A) Tracing
   B) Version Control
   C) Metrics
   D) Logging
2. **What is the primary purpose of metrics in observability?**
   A) To provide detailed event logs
   B) To manage code version history
   C) To track system health and performance
   D) To capture user input in real-time
3. **Distributed system failures are best diagnosed using which observability pillar?**
   A) Monitoring
   B) Debugging
   C) Logging
   D) Tracing
4. **Which of the following best describes a Key Performance Indicator (KPI)?**
   A) A performance metric that defines system goals
   B) An AI-driven log parser
   C) A list of static code analysis rules
   D) A random sample of system events
5. **What is the role of real-time dashboards in monitoring?**
   A) To deploy cloud-native applications
   B) To test user interface responsiveness
   C) To visualize and track system health metrics
   D) To execute database queries

6. **Why is categorizing incidents as critical or non-critical important?**
   A) To prioritize incident response and resource allocation
   B) To assign code reviewers efficiently
   C) To document software licenses
   D) To reduce deployment times
7. **What is one major benefit of automating incident response?**
   A) Prevents the need for monitoring tools
   B) Eliminates the need for any human oversight
   C) Reduces costs by removing infrastructure
   D) Speeds up resolution time and reduces manual effort
8. **How much time do Site Reliability Engineers (SREs) typically spend on on-call duties?**
   A) One week per year
   B) Only during outages
   C) 25% of their time, such as one week per month
   D) 50% of their total work time
9. **What is a key feature of a blameless postmortem?**
   A) Avoiding documentation
   B) Encouraging honest analysis without fear of punishment
   C) Assigning responsibility to individuals
   D) Only focusing on system uptime

10. Why is building a culture of continuous improvement important in incident management?

A) To eliminate the need for postmortems
B) To reduce the need for training
C) To avoid spending on observability tools
D) To drive innovation and enhance reliability over time

Answers

1.	2.	3.	4.	5.	6.	7.	8.	9.	10.
B	C	D	A	C	A	D	C	B	D

CHAPTER 04

Automation and Infrastructure as Code (IaC)

4.1 The Role of Automation in Site Reliability Engineering

The use of technology in businesses is critical due to the fast and ever-changing digital world of today where businesses are using the same to drive their operations, provide a seamless customer experience and remain competitive. With organizations growing and adopting more and more complex infrastructures, reliable and resilient, and high performing systems are essential. Site Reliability Engineering (SRE) is exactly what's needed in this place. SRE is relatively young but rapidly growing in popularity because it is a different kind of way to ensure that critical systems and applications keep running, keep running faultlessly, and remain secure (Efraim Turban, Carol Pollard, 2021).

This practice is about site reliability engineers (SREs). Each of them is responsible for wide range of tasks which altogether guarantee the reliability and performance of the digital services. They are also responsible for continuous monitoring of system's health, incident response & postmortem analysis, infrastructure optimization & capacity planning, performance tuning, automated deployments & recoveries, etc. This role puts the work of software engineering into the work of system administration and helps bridge the gap between them to build software systems that are highly scalable and highly reliable.

The most pressing SRE challenge they feel is managing their never ending heterogeneous infrastructure that lives across one, multiple or all of on-premise data centers, hybrid cloud, and multi cloud platforms. Secondly, they operate under resource constraints and are subject to the pressure of minimizing downtime, fast recovery from failures, and provide high quality services while keeping innovation speed. And these demands have a peculiar requirement of a careful balance between reliability and velocity.

Therefore, SREs have increasingly embraced automation and advanced solutions of Artificial Intelligence for IT Operations (AIOps) to tackle these challenges. The automation takes away the error of humans, speeds up the repetitive task like deployments and scaling and lets the team work on strategic improvement rather than firefighting. In contrast with AIOps uses machine learning, data analytics and pattern recognition to enhance visibility, predict incidents to happen, and intelligently filter out alerts to reduce noise and have shorter incident resolution times.

By using these tools and practices, SREs can not only keep the systems running and performing up to par but also contributes quite immensely to improving the business agility and customer satisfaction. On the other hand, they are at the heart of allocating future opportunities for reliable computing in a digital world where the lack of available digital systems is not acceptable.

Benefits of automation in SRE

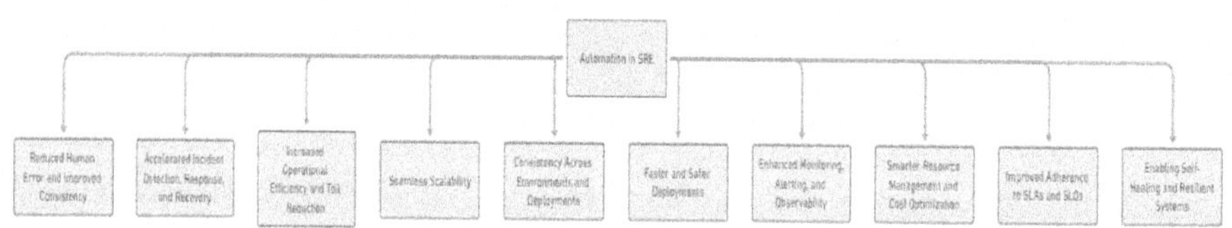

Figure 4.1: Benefits of automation in SRE.

Source: (Self-generated)

Automation in Site Reliability Engineering is not just a convenience, but a constant, necessary principle that modernizes the development of robust, scalable and highly available systems. Automation enables SREs to reliably meet reliability goals without being toiled and human error (Smith, 2021).

The expanded key benefits of automation in SRE are as follows:

1. Reduced Human Error and Improved Consistency

Reducing human error is one of the main contributors to automation in SRE. Mistakes are generally made in manual interventions especially high-pressure situations such as incident response, which can cause service outages and data loss. Automation guarantees tasks follow exact defined course of action every time it is run, an action determined by preconfigured logical rules. This is particularly important in deployment, infrastructure provisioning, system update, failover type of tasks, where even a small error can lead to a huge impact.

2. Accelerated Incident Detection, Response, and Recovery

Incident management workflows are very much enhanced by Automation. Systems are able to monitor, or detect anomalies, thresholds breach, or failure in real time through automated monitoring and alerting. Additionally, automated remediation workflows can automatically do such things as restarting services, scaling up resources, isolating failing components or triggering rollbacks before human involvement is required. So, the Mean Time to Detect (MTTD) and Mean Time to Recover (MTTR) are decreased, two critical metrics that keep the system reliable and running.

3. Increased Operational Efficiency and Toil Reduction

The toil involved in the SRE teams is sometimes manual, repetitive work that provides no long-term value. This is addressed by automation, offloading routine acts like server provisioning, disk cleaning, log rotation, back-ups and patching, etc. In SRE by reducing toil, SREs get more time and energy to focus on running strategic endeavors, like refining the system architecture, tightening the system performance, and creating tools to make the reliability even better.

4. Seamless Scalability

In modern applications, it is no longer acceptable for at scale to take a significant amount of time. Elastic scaling is enabled via automation of provisioning or deprovisioning of resources depending on 'real time' metrics like current CPU usage, current memory pressure or request throughput. It makes sure that systems can burst traffic and shrink it back to save costs without any manual intervention.

5. Consistency Across Environments and Deployments

Automating teams allows them to standardize the workflows and the infrastructure configurations through techniques such as Infrastructure as Code (IaC). This makes sure that environments, development, staging, and production match and are reproducible. Furthermore, automated deployment pipelines assure that builds are tested, validated and deployed in the controlled and reliable way which reduces the possibility for the environment specific issues and the configuration drift.

6. Faster and Safer Deployments

Automation brings the Continuous Integration and Continuous Deployment (CI/CD) practices and streamlines the whole software delivery lifecycle. With the changes, all can be tested, built, and deployed in the same pipeline with integrated checks decreasing its deployment time down to minutes instead of hours or days. Automation ensures that deployment failures can be rolled back quickly, safely to a known good state and safely minimizes disruption into users and reduces risk.

7. Enhanced Monitoring, Alerting, and Observability

Automation is also used by modern observability platforms to collect and process huge amounts of telemetry data, logs, metrics and traces from all over the system. Such data can be processed in real time to find patterns, generate insights and alert on context aware basis with the help of AIOps and automation. It prevents alert fatigue and allows on call engineers to receive actionable information only rather than wasting time responding to unrelated issues, resulting in faster response time and better engineers' mental health.

8. Smarter Resource Management and Cost Optimization

SREs can automate provisioning, resource scaling, and lifecycle management of infrastructure, so that the provisioning, scaling and lifecycle of infrastructure will be optimized and the overhead is simplified to what is required. Resource consumption tool can keep track of resource consumption trends and make decisions based on the data on whether idle instances should be shut down, if workloads should be right sized, or if jobs should be scheduled during off peak hours. Better utilization of the infrastructure and major cost savings, especially with cloud environments, results from this.

9. Improved Adherence to SLAs and SLOs

Automation helps meeting the reliability goals defined by Service Level Agreements (SLA) & Service Level Objectives (SLO) by keeping up the time required to respond as well up time and system availability. It assures that performance thresholds are maintained and breaches are discovered and taken care of rapidly. Unfortunately, this does not guarantee reliability targets and prevent contractual or reputational damage resulting from SLA failure, which brings us to the automation side.

10. Enabling Self-Healing and Resilient Systems

Modularization as the basis of automation is one of the most powerful application in the field of self-healing systems (systems with the inherent ability to identify and repair themselves when a fault occurs). Systems are able to take corporate action when they respond to violations by means of automated runbooks, scripts and workflows. To name an example, if a service goes down, it will automatically be restarted; if a server is unresponsive, traffic can be rerouted. It does this and reduces downtime, which makes the environment more resilient with a production environment that is more robust.

In short, while automation is a tool for SRE, it is also a philosophy to scale the systems, reliability and performance of modern digital systems. SRE teams can automate critical operational tasks to have more stable environments, to respond quicker to incidents, to save more money and work on continuous improvement. In cases of increasing complexity of systems, the use of intelligent automation will increase, with it becoming a key element of successful reliability engineering practices.

4.1.1 Automating Repetitive Operational Tasks

Site Reliability Engineering (SRE) is automation of repetitive tasks in the operation that takes less human effort and more reliable systems. They are manual, repetitive, and don't really require much of deep human judgment, and they are called "toil." Some of these such as log rotation, service restarts, patch deployments and user account provisioning. However, SREs are freed up by automating such tasks for strategic engineering work such as improving system performance, scalability, observability, etc.

The operational processes are made consistent and accurate with automation. Human execution has errors (and mistakes) that are large enough to occur under stress and scale, automated systems are much smaller errors every time because of defined steps. Imagine for instance Infrastructure as Code (IaC) tools like Terraform and Ansible that are examples of a commitment towards building reproducible environments in producing results at a perfect level of comfort and safety, without any configuration drifts and all deployments being consistent and reliable. In CI/CD pipelines, build, test and deployment are automated even further to eliminate the risk of putting bugs into production environments when delivering software even further.

Process automations also help in improving incident response and self-healing systems. In doing so, such monitoring systems can detect anomalies and automatically trigger corrective actions with the help of tools like Prometheus, Grafana, run books and others. This can involve restarting failed service, bringing in new instances for peak traffic, handling requests to healthy servers – lowering Mean Time to Recovery (MTTR) and overall service availability. It also means that the system is scheduled for health checks, security updates and capacity scaling proactively.

It goes beyond the technical efficiency of automation and also helps improve the engineer's experience and scale the organization. It reduces cognitive load to the team, increases team morale and helps organizations scale infrastructure without scaling head count in proportion. Lower cost is the natural result of less manual labor and less downtime. Ultimately the goal of automation is to save time, but automation of repetitive tasks helps enable building more reliable and scalable, and more resilient systems to handle both modern demand, as well as give teams the chance to deliver on innovation and value that is connected to improvement.

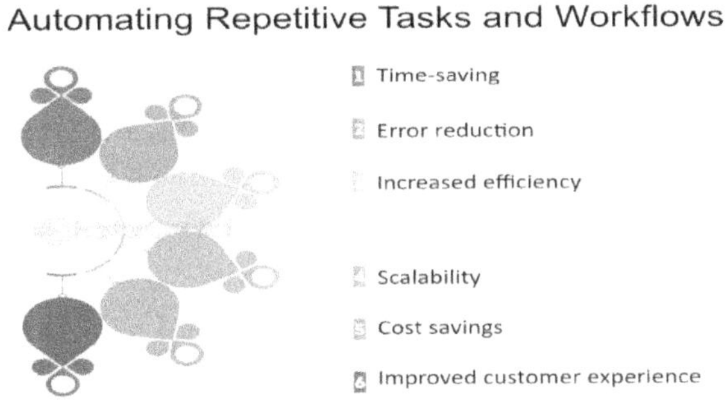

Automating Repetitive Tasks and Workflows

Time-saving
Error reduction
Increased efficiency
Scalability
Cost savings
Improved customer experience

Figure 4.2: Benefits of Automating Repetitive Tasks and Workflows in SRE.

Source: (FasterCapital, 2020)

1. **Time-Saving:** Automating systems dramatically cuts the time spent on man's work. Automation scripts or tools can execute all such operations instantly and concurrently. This frees up the Site Reliability Engineers to redirect their focus toward high impact engineering topics like system optimization or new feature development.
2. **Error Reduction:** This type of interventions are prone to human error due to manual and are usually not more than human error like during a time of high stress, such as incident response. All in all it allows to automate the execution of a task in a predictable manner with a standardized approach to minimize the risks of messed configurations or overlooked steps. This makes the system more stable and less prone to unreliable outcomes in production environments.
3. **Increased Efficiency:** The workflows will become more automated and the operational workflows will become streamlined and faster to complete where there is less oversight. For instance, CI/CD pipelines speed up the release velocity and reduce bottlenecks when releasing code by handling, like, building, testing and then deploying the code at various stages.
4. **Scalability:** With increasing size and complexity, the manual operations become unfeasible. Automation allows horizontal scaling because it allows managing new workloads without having to increase manpower in proportion. Kubernetes provides this type of scaling out of services automatically based on traffic, or CPU usage, so the systems are always fluidly scaling based off of demand.
5. **Cost Savings:** Both saves on manual effort and downtime and is very cost saving. Automating infrastructure management and dynamic resource provisioning reduces over provisioning and optimized hardware utilization with less operational overhead (bringing not just technical advantage but financial one too).
6. **Improved Customer Experience:** Services are made available, responsive, and trustworthy in event of traffic spikes, partial

failures or none at all with automation. Load balancing, incident response automation and self-healing mechanisms ensure that the user experience is continuous, because such is the competitive digital environment of today.

SRE automation helps to save time, reduce errors, make things more efficient, and to run in a scalable, inexpensive manner. This guarantees great customer experience through responsive, self-healing systems.

4.1.2 The Challenges of Over-Automation

Automation is definitely a great boon to Site Reliability Engineering (SRE), but over automation comes with its own set of challenges. Its one major concern is too much reliance on automated systems in which engineers lose contact with the systems they're in charge of. It can result in little or no situational awareness during incidents, whereby the cause of events are not clear and creative responses are not available when automated processes fail (Pethuru Raj Chelliah, Shreyash Naithani, 2018).

There is another key issue: maintaining these automated systems themselves are complex. The more it scales, the more scripts and triggers adding layers, and then dependencies that will turn thorns over time. Without proper documenting and maintaining it can result in an unexpected failure or "automation fatigue", i.e., the costs to managing automation exceed the benefits.

The challenge of automation drift exists, where automated processes change ways from what they were intended to behave, given changes to the environment or inattention to version control. Regular audits and governance are needed to keep automated systems fresh, preventing the way they grow unintended be bugs or misaligned with organizational goals.

Finally, over automation can stop innovation and critical thinking. In case teams are too dependent on the predefined scripts or tools, they might ignore the possibility to refine the system design or automate processes manually. SRE teams need to have the right balance between automation and active learning, human judgment, and the practice of continual improvement.

4.2 Infrastructure as Code (IaC): Tools and Best Practices

Infrastructure as Code, also referred to as 'IaC,' is a revolutionary paradigm in today's time of infrastructure management. The term describes itself, it is a practice of provisioning and managing computing infrastructure through machine readable definition files instead of manual physical configurations or traditional setup methods. The benefit of this approach is it treats infrastructure like application source code by being precise, having version control, automating, and being repeatable when working with infrastructure. It signifies that it's taking the discipline and practices that the development teams have to software code apply to infrastructure as well.

The process of managing infrastructure used to be a very manual affair. Scripts and graphical interfaces were used to ran system administrators into servers, schedule some configuration, patch systems, or deploy environments. The process was quite time consuming as well as prone to human error. By definition, Infrastructure as Code describes configuration files as being the representation of the intended state of a set of infrastructure components such as servers, load balancers, databases and network settings.

IaC is simple and fast to provision resources and is therefore one of the greatest advantages. The DevOps professionals (SREs) need not execute hundreds of disparate configuration steps to change infrastructure environments anymore. Using IaC, the same result can be achieved with a single command or script. Pre-defined infrastructure scripts can also be used to spin up a new development environment or scale production resources so that they can meet increased demand, which can all be done instantly. The same script can also be run at the end of the day to downsize idle resources, resulting in a more efficient and cost-effective method of operation.

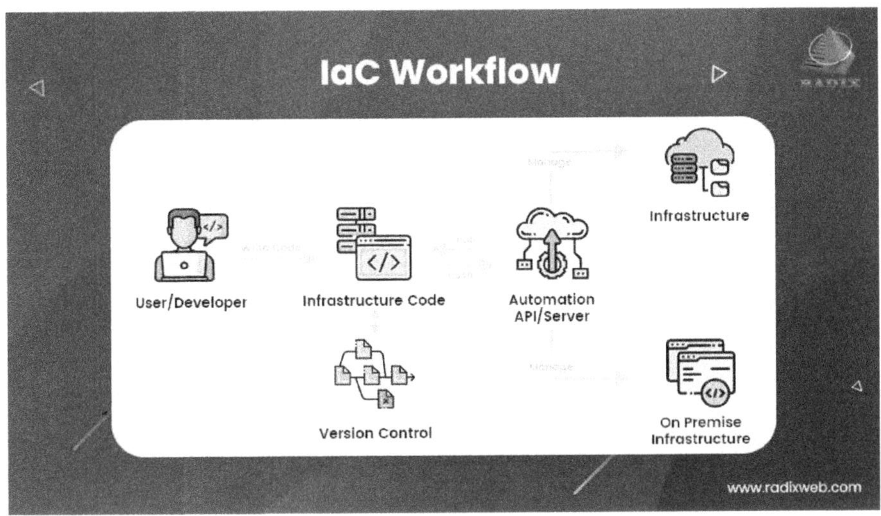

Figure 4.3: Infrastructure as Code (IaC) Workflow with Automation and Version Control.

Source: (Mistry, 2024)

IaC also introduces software development best practices to infrastructure management. Infrastructure code can be stored in versions control systems, tracked, reviewed history, and even automated testing and deployment pipelines can be implemented. This also provides consistency and reliability as well as supports collaboration between teams. IaC when applied right helps to create a culture of transparency, repeatability, and continuous improvement, which is a perfect match to the principles of agile development and the DevOps methodologies.

While ITC is a concept that gives great flexibility and inline management to Infrastructure, its implementation is infeasible as organizations grow bigger and embrace cloud native architectures. It simplifies dramatically the complexity of management of hybrid or multi cloud environments and provides the basis to develop scalable, resilient and reproducible systems. In a nutshell, IaC turns infrastructure into a software component, which is testable, programmable and reliable part of the software delivery lifecycle.

4.2.1 Introduction to Terraform, Ansible, and CloudFormation

1. Terraform

HashiCorp's Terraform is a popular IaC (Infrastructure as Code tool) that you can use to deploy the infrastructure across most of the cloud providers like AWS, Azure, Google Cloud Platform, etc. This cloud-agnostic design makes Terraform powerful because it allows users to adopt a single approach to manage multi-cloud or even hybrid infrastructures with ease.

Terraform uses a declarative language (called HCL, or HashiCorp Configuration Language) so that through a series of human readable, simples code it can define the final state of infrastructure for free. After defining, terraform computes the dependency graph and computes the least amount of effort to achieve the defined state. Terraform uses its state management to track all the managed resources and makes sure changes are applied incrementally and predictably. Due to this it is very suitable to manage very complex infrastructure at scale with little to no risk of configuration drift or manual error. Reusability and collaboration of teams are encouraged by this design of Terraform's modular design, making infrastructure code consistent and version controlled.

Advantages:

- **Platform Agnosticism**: Heterogeneous infrastructure can be deployed to multiple clouds, thanks to it supporting different cloud providers for its orchestration.
- **Declarative Syntax (HCL)**: More reproducible and predictable in that a specification for the state is specified rather than procedural steps.
- **Immutable Infrastructure Model**: It encourages infrastructure rector over in place update and helps in reducing the configuration drift and improves environmental consistency.

- **State Management**: Keeps a persistent state file and allows incremental changes that help avoid unwanted side effects during the updates.

Disadvantages:

- **State File Vulnerability**: Centralizing the state file into one point of failure can make the system vulnerable to failure, and it often requires secure backend management to be available and consistent.
- **Complex Debugging**: In cases where failures are inter-resource dependent, troubleshooting can be non-trivial without a lot of intuitive feedback.
- **Learning Curve**: Newcomers unfamiliar to declarative paradigms may find the use of HCL and the abstract resource dependency model to be difficult.

2. Ansible

Red Hat is a company that develops the Ansible automation engine, which is an open-source automation engine that pays particular attention to the management of configuration, deployment of applications, and the orchestration of infrastructure. Ansible is one of those applications that operates agentless architecture. Unlike other tools that require the installation of daemons or agents on managed nodes, Ansible communicates directly with machines via SSH (or WinRM for Windows) making it very easy to setup and secure.

Ansible uses YAML based playbooks that define a series of tasks to be done on the target systems in procedural fashion. This is because operations are done in a fine grained fashion, which makes it suitable for controlling post provisioning configurations like installing packages, configuring services, deploying applications, etc. Day to day tasks, configuration consistency and reducing manual intervention are all widely liked by DevOps teams of automating their process using the simplicity, readability and extensibility provided by Ansible. Ansible also supports idempotency (the same result is generated on repeated execution), which leads to system stability and predictability.

Advantages:

- **Agentless Architecture**: It uses standard protocols like SSH or WinRM so it has little overhead and deployment in different environments is easy.
- **Procedural and Idempotent Execution**: It provides a step-wise control over the configurations while ensuring repeatable and stable outcomes.
- **Extensive Module Library**: It provides a huge collection of modules for the OS level configurations, cloud services, applications, and containers.
- **Readability (YAML)**: Human readable syntax helps collaboration across the development, operations and security teams.

Disadvantages:

- **Scalability Limitations**: Execution performance of the system might be degraded on very large infrastructures where task execution is sequential and no parallelism is native.
- **Limited Native Infrastructure Provisioning**: It is not as suitable for provisioning and provisioning low level resources of infrastructure as any other tool like Terraform.
- **Error Handling Constraints**: Risks are incurred to the procedural nature of the system, and to the lack of rollback mechanisms, but only if task failures or partial deployments are allowed.

3. AWS CloudFormation

Amazon's proprietary Infrastructure as Code service that enables developers and system administrators to provision AWS resources through a JSON or YAML template. As opposed to Terraform, which is a platform agnostic tool, CloudFormation is Github, but limited to AWS and has native integration with the entire suite of AWS services.

With CloudFormation users can create complex architectures such as compute instances, storage volumes, a databases, and more; all in the same template that has been version controlled. From there, CloudFormation adds all kinds of powerful features for nested stacks, change sets, and world; things which make working with infrastructure safe and simple for teams. Infrastructure changes are predictable; thus, rollbacks are possible when things fail and it have consistent environments between development, staging and production. By leveraging CloudFormation, teams committed to the cloud can use a robust, secure, and deeply integrated manner to both manage and scale cloud environments that keep their infrastructure under infrastructure governance policies.

Advantages:

- **Tight AWS Integration**: Real time feature parity and access to new service capabilities with native compatibility to all AWS services.
- **Infrastructure Governance**: CloudTrail integration provides additional support for drift detection, resource policies, as well as audit trails.
- **Transactional Resource Management**: It uses a rollback on failure mechanism to do atomic deployments and prevent incomplete states.
- **Support for Nested Stacks**: It encourages modular and hierarchical infrastructure design and will make the template more maintainable and reusable.

Disadvantages:

- **AWS Exclusivity**: Potentially locking it into one vendor, not applicable in the case of multi cloud strategy or hybrid cloud environment.
- **Verbose Templates**: The definitions of JSON/YAML can become very large and hard to manage without the help of external templating engines or abstractions.

- **Slower Iteration Speed**: Updates to the stack and change sets can be time consuming, and especially in a largescale environment with many inter dependencies.

4.2.2 Managing Infrastructure as Code at Scale

As organizations expand their cloud-native environments, managing infrastructure efficiently at scale becomes a formidable challenge. Infrastructure as Code (IaC) enables teams to provision and configure systems through declarative or procedural scripts, bringing automation and repeatability to traditionally manual processes. However, as infrastructure footprints grow, so does the complexity of maintaining and coordinating these systems. Scaling IaC practices requires thoughtful design, standardization, and robust tooling to ensure reliability, security, and efficiency (Chinamanagonda, 2019).

One of the foundational principles for managing IaC at scale is the adoption of modular and reusable code. Rather than writing monolithic templates, infrastructure should be broken down into small, composable modules that can represent specific services, components, or environments. This modularization allows teams to reuse tested patterns across different parts of the organization while maintaining consistency and reducing duplication. For example, a standard module for provisioning a database cluster or a network configuration can be shared across multiple projects, enabling centralized governance and improved collaboration.

Version control plays a critical role in scaling IaC by introducing transparency and accountability into infrastructure changes. Storing IaC definitions in Git repositories allows teams to track changes over time, implement peer reviews, and roll back to previous configurations when needed. To further separate concerns and prevent configuration drift between environments, organizations typically maintain distinct repositories or branches for development, staging, and production deployments. GitOps workflows have become increasingly popular, as they extend these principles to automate the application of infrastructure changes using pull requests and continuous integration/continuous deployment (CI/CD) pipelines.

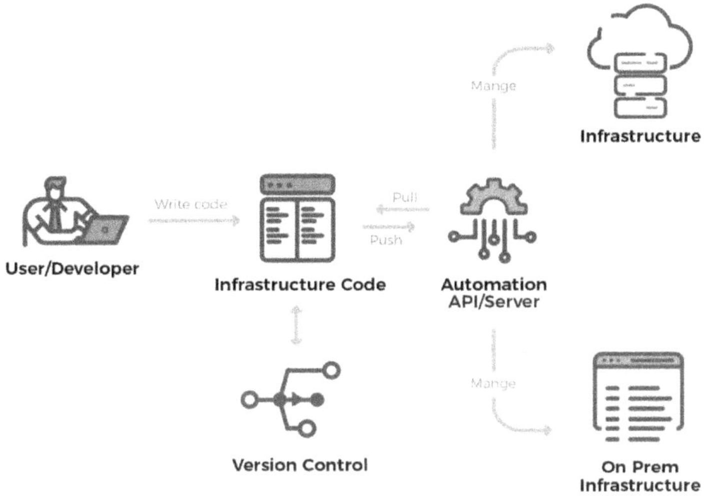

Figure 4.4: End-to-End Infrastructure as Code Workflow – From Developer to Managed Infrastructure.

Source: (Kunal, 2023)

Managing infrastructure state across large teams and projects presents another layer of complexity. Tools like Terraform require a state file that maps infrastructure resources to their configurations. At scale, this state must be stored remotely and securely to support collaboration. Remote backends such as AWS S3 with DynamoDB for state locking or Terraform Cloud provide a centralized and consistent way to manage infrastructure state while preventing conflicting changes. Access control mechanisms should also be enforced to limit modifications to sensitive or critical resources.

Security and compliance are paramount in large-scale IaC environments. To minimize risks and ensure adherence to governance policies, many organizations incorporate policy-as-code tools into their deployment pipelines. Solutions like Open Policy Agent (OPA) or Sentinel can automatically validate infrastructure definitions against security baselines, compliance standards, or cost controls. Moreover, integrating secret management systems, enforcing role-based access controls (RBAC), and maintaining audit trails all contribute to a more secure and accountable infrastructure environment.

Testing and validation of IaC become increasingly important as the number of resources grows. Automated testing tools such as Terratest or KitchenCI can verify that infrastructure behaves as expected before deployment, reducing the chances of misconfigurations or service outages. These tests can simulate different conditions, validate outputs, and ensure that infrastructure code continues to meet requirements even as dependencies evolve.

Over time, discrepancies can arise between the declared infrastructure in code and the actual state in the cloud environment—a phenomenon known as configuration drift. Tools like Terraform's plan command or AWS Config allow teams to detect and correct drift, preserving consistency between the infrastructure definition and the real-world environment. Continuous monitoring, combined with alerts for unauthorized changes, helps ensure long-term stability and compliance.

Finally, scalability and cost-efficiency are closely linked in large infrastructure deployments. IaC tools allow teams to automate the scaling of resources based on demand, integrate with monitoring platforms to optimize utilization, and prevent over-provisioning. By enabling teams to dynamically allocate, deallocate, or resize infrastructure based on performance metrics or billing insights, IaC contributes to both technical and financial sustainability.

4.2.3 Version Control and IaC Best Practices

One of the important reasons why version control is important for managing Infrastructure as Code (IaC) is that it offers a secure and traceable mechanism of understanding changes, teamwork among projects, or consistency in configuration between environments. To that end, with tools like Git itcan store all infrastructure definitions, scripts, and templates as nothing but a centralized repository and collaborate, review changes, as well as rollback to previous states if needed. On top of that, this makes infrastructure management practices match what software development already does (Donaldson et al., 2020).

To make sure that the system is stable, scalable and secure, it is important to implement best practices in version control for IaC. Infra code is already equal parts of the foundational practices itoutline here like Commit messages, Atomic commits and use Feature Branches for code that is specific and good practice. Pull requests and performing peer reviews helps in catching configuration mistakes early and hence improves overall quality and lessens the possibility of introducing major issues in production environments. Further, tagging releases and maintaining versioned environments makes the entire process by looking more reproducible and roll they can do back in the event of error.

However, one of crucial best practice is modularization, breaking down the code to an infrastructure into reusable smaller modules that identify the logical components, e.g., networking, compute or storage. This also helps maintain the code and reuse it for projects, and allows for independent testing and updating of the code without affecting the whole infrastructure stack. The typical way of preserving secrets and delicate variables while continuing to be transparent in configuration management is to deploy modules and store specifically configured files and sensitive data in safe systems like Vault or SOPS. By doing so, itensure that confidential information remains encrypted and protected, while the entire infrastructure is at the same time auditable and manageable.

Overall, combining version control with proper adhere to best practices in IaC is exactly what helps infrastructure to be not only automated but also predictable, secure, and manageable at scale. It allows teams to work faster, to detect and fix issues quicker, and to maintain a healthy record of all infrastructure related changes for improving system reliability and resilience of modern cloud native systems.

4.3 Configuration Management with Ansible, Terraform, and Kubernetes

Today, Terraform, Ansible and Kubernetes together form a very strong and strategic combination for configuration management in today's fast paced and dynamic infrastructure environments. As opposed to operating in silos, this layer of tools makes up an open, seamless, layered automation ecosystem that fully address infrastructure and application deliver hardware lifecycle, including provisioning and configuration, orchestration and runtime management (Vikas Grover, Ishu Verma, 2023).

The initial phase of this integrated workflow is Terraform. It is used to provision and define infrastructure in a consistent, declarative way across different providers or on premises. Terraform makes sure whether networking components are being configured, setting up virtual machines, managed Kubernetes clusters, all the infrastructure is reproducible and version controlled. This helps organizations to scale their infrastructure efficiently and keeping consistency in development, staging and production environments.

After that, Ansible comes in to configure the systems in the newly created infrastructure. Terraform is about "what" while Ansible is about 'how', dealing with system level services deployment, installing software & handling post provision steps such as configuring environment, installing required software, security patches deployment. The second layer of automation between infrastructure and applications bridges the gap between configuration of every component to meet operational standards.

This stack is then orchestrated and executed on top of it by Kubernetes. When the infrastructure is set up and the environments are launched, containers deployed and managed using Kubernetes at scale. It offers strong self-healing, self-balancing, auto scaling capabilities and resilient and responsive application delivery. Secrets and configuration data that can be injected through Ansible or managed through Vault/SOPS can be securely added to Kubernetes deployments and remain automated and compliant.

The integration of this approach makes it streamlined while helping leaders know best DevOps practices and Site Reliability Engineering. It always promotes modularity, repeatability, and transparency from the whole deployment pipeline. Each layer is defined, automated and versioned, meaning the risk of changes in infrastructure or application configuration is minimal, as it can be done with little to no risk of affecting other layers. The agility to innovate faster with higher reliability, availability and security is therefore enabled to teams.

However, by using Terraform, Ansible, and Kubernetes together, an organization can have the capability to deal with the complexity at scale, minimize operational overhead and design robust systems that are ready to scale on continuously and change the business need.

4.3.1 Automating Configuration Changes Across Environments

Automating configuration changes in a number of environments is fundamental to modern infrastructure management, though especially for Site Reliability Engineering (SRE) and DevOps practices. Updating manually brings with it the risk of human error, and makes it less likely that the environments will be consistent between development, staging, and production. For this challenge, automation tools such as Ansible, Terraform and Kubernetes are all combined to use them to make that configuration changes are reliable, repeatable, and scalable (Dekker, 2017).

Teams are able to define the infrastructure and environment specific settings of infrastructure using Infrastructure as Code (IaC) tools such as Terraform through version-controlled files. This allows us to use the same baseline infrastructure on several environments with slight variations under control through variables and modules. For example, terraform modules can be reused across environments and have different configurations (instance size, database credentials) injected during their use with environment specific variable files. This avoids duplication and guarantees that each environment is in line with the set standards.

Infrastructure that is provisioned will then be configured at each host using Ansible or other configuration management tools. Idempotence of ansible playbooks means that ansible playbooks can be run multiple times and the result will not change beyond the first application. Parameterizing playbooks is also possible so that they can be used on different environments and that configuration drift is avoided, and systems remain uniform across all environments. Integrating Ansible playbooks with a CI/CD pipeline helps organizations to automate deploy of configuration changes in a controlled and auditable way. The playbooks in the pipeline can send automatically or triggered by events, i.e. ones that are triggered by code commits or approved pull requests. This approach not only ensures that there will be no errors in this and also makes the updates to updates consistent, repeatable, and in trace a way.

In Kubernetes based environment, configuration changed is managed declaratively using YAML manifests and Kubernetes APIs. Such as; ConfigMaps and Secrets are used to inject configuration with environment-specific injection without modifying the code of the application. Any change in Git repository configuration is synced automatically in the target environment through GitOps workflows using tools such as Argo CD or Flux. This helps changes to be reviewed, tested, and promoted through environments in a structured and transparent way.

In summary, automating configuration changes on their environment make the team able to be more agile, quick to deploy and more reliable. It helps eradicate manual inconsistencies, support rollback and audit capabilities, and keep engineering teams at ease with their infrastructure, however complex or the number of environments it might entail. It provides the foundation for the robust, scalable and error resistant system administration.

4.3.2 Using Kubernetes for Scalable Deployments

Kubernetes is becoming a foundation technology for running and managing the containerized applications at scale. It offers automated, resilient and scalable platform that is effective in supporting deployments across various environments in modern cloud native and Site Reliability Engineering (SRE) practices. This means that Kubernetes abstracts the complexities of underlying infrastructure and declutters it for vertical and horizontal scaling of the applications on demand (Poulton, 2023).

Horizontal scalability in Kubernetes means that applications can be scaled in or out in real time as per the real time demand. Built-in tools such as Horizontal Pod Autoscaler (HPA), help us in managing this process and use of resource efficiently and provide consistent performance.The HPA watches metrics like CPU and memory usage or custom application level indicators and change the number of running pods. All of this means that if the amount of system loads that need to be delivered changes over time, the dynamic scaling will guarantee the best performance without any intervention whatsoever when the amount of work to do changes. The Vertical Pod Autoscaler (VPA) can also scale resource allocations for individual pods to tune memory and CPU limits as they are needed.

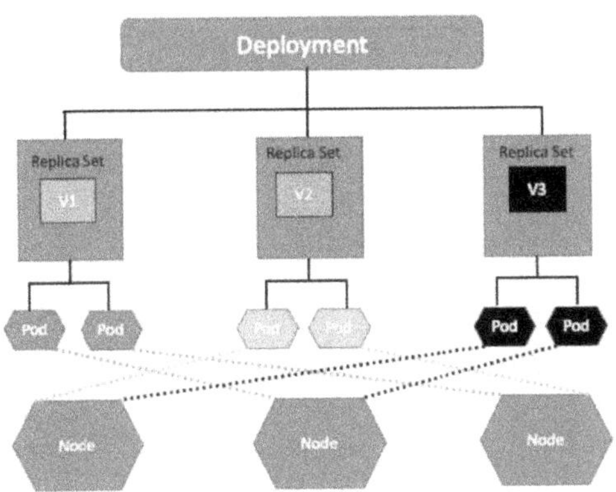

Figure 4.5: Kubernetes deployment for cluster auto-scaling.

Source: (Zeydan et al., 2022)

These are critical ways of delivering a scalable and safe application and Kubernetes also facilitate rolling updates, canary deployment. These features allow updates of applications be scripted where downtime is minimized and disruption risk minimized. In this way, users have access to services while services are being deployed, tested, and validated in real world traffic scenarios. Kubernetes also offers rollback mechanisms, which can revert to a previous stable version, in case of a failed deployment, an important feature of high reliability.

Furthermore, Kubernetes can be integrated with CI/CD pipelines and GitOps workflows to help scale deployments on a large scale across all of the environments or regions. These integrations that autosync the releases and have automated deployment reduce manual release overhead. Kustomize, Helm and Argo CD showed up as tools for teams to manage complex application configurations at scale and to keep their configurations in sync across different clusters (Dev, Staging and Prod) and build services on top of services.

To sum it up conclusively, by its robust orchestration, Kubernetes can deliver scalable, self healing, and highly available services in organizations. Dynamic resource management, declarative deployments and automated scaling are ensured to allow it to grow with user demand while reducing service reliability or operational efficiency due to dynamic management of resources. Indeed, Kubernetes is not a container orchestrator, though it is a key enabler of scalability and resilience of modern digital infrastructure.

4.3.3 Policy Enforcement and Security in Configuration Management

Policy enforcement and security are both pillars of the operational integrity, the regulatory compliance, and the robustness of the system in the context of configuration management, especially in large scale and cloud native environments. As infrastructure and applications are being more and more defined and managed through code, it needs rules and security constraints defined and enforced programmatically to shut these rules and security constraints out of propagating environments (Bulmer, 2024).

Policy enforcement is the automated application of rules that control how resources can be defined, deployed, as well as managed. This is done in tools like Terraform, usually through frameworks like Sentinel or Open Policy Agent (OPA) for writing one or more custom policies in code and validate them before infrastructure changes are made. They can be used to restrict certain instance types, enforce tag usage, or deny accessing storage buckets by the public. Tools such as Kyverno and Gatekeeper allow Kubernetes to enforce policies on resource definitions to either monitor and validate resource definitions to ensure that deployments adhere to an organizational policy such as, prohibiting privileged containers, or enforces required network policies.

In configuration management, security is about the fact that sensitive data like secrets, credentials and tokens, are never exposed as plain text, and are always handled through secure, encrypted means. IaC secrets are securely managed via tools such as HashiCorp Vault, Mozilla SOPS, AWS Secrets Manager and integrated into IaC workflows without going over confidentiality. Secrets should be referenced dynamically at runtime in the configuration files, and access to such secrets should also be role based and audited.

Further, the secure configuration management includes least privilege access, immutable infrastructure principle and regular security audits. To lock down automation pipelines, they have to be locked down and only authorized changes are deployed, and signed commits, pull request approvals, and policy checks are enforced as part of the CI/CD process on version control systems. Integration of scanning tools can also be used to find hardcoded secrets or insecure configurations before they hit production.

The technical mechanisms are not sufficient, instead they are also about a culture of compliance and accountability through policy enforcement and security. It needs to train teams to do the right thing in terms of configuration, code reviews, credential management etc. Guardrails need to be automated and let developers move fast with the safety intact.

In the end, good policy enforcement and strong security practices in Configuration Management avoid downtime, reduce the attack surface and keep configuration and infrastructure on point with business and regulatory requirements. But these safeguards become indispensable as the complexity of infrastructure rises in order for the continuity and control to remain.

4.4 CI/CD Pipelines: Automating Software Deployment for Reliability

Today, organizations are forced to bring software out quickly, often, and with a high degree of reliability. Traditional software development and deployment approaches—ranging from manual and fragmented to imprecise and fragmented—are no longer adequate for agile development cycle, dynamic customer expectations and competitive market. To respond to this issue, Continuous Integration and Continuous Deployment (CI/CD) pipelines are been used as a modern DevOps practices' basic part. Automated workflows such as these reduce the time taken from the code commit to the production release of software, improve the software quality, increase software delivery speeds, and ensure system stability (Highsmith, 2002).

Continuous Integration (CI): The Foundation of Automated Collaboration

It is a development practice in which the team members integrate their code changes into a shared repository several times a day. Automated build processes and test suites allow for errors to be found as early as possible, and each integration is automatically verified in turn. The objective is to get problems in the integration, improve the code quality, and have faster development pace.

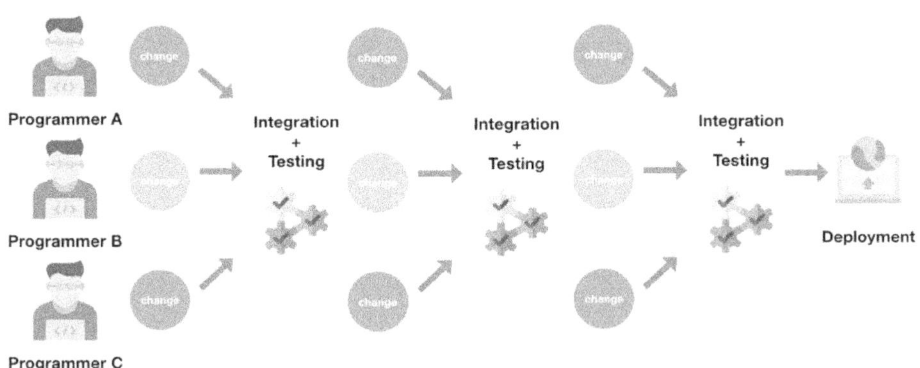

Figure 4.6: Continuous Integration Workflow with Multiple Developers.

Source: (Sky, 2020)

The accompanying visual shows how individual programmers (Programmer A, B and C) develop features or fix bugs without any collaboration. A central codebase is created, in which each change is represented by a 'change' node and integrated. The changes are automated for static code analysis, unit testing and build compilation. If done right, CI becomes a well-defined and predictable process for integrating code, thereby eliminating the 'merge hell' which comes with long lived feature branches.

Benefits of CI include:

- **Faster detection of defects**: Due to the faster release cycles, there is faster feedback and innovation that can be delivered.
- **Increased transparency**: Deployment processes are not dependent on the individual developer and operations teams.
- **Reduced integration issues**: Smaller changes are much easier to test and easier to go back and rollback if any problems occur. Lesser chance of deployment failure.
- **Improved team collaboration**: More reliable and faster transportation of new features, enhancements and bug fixes for the customers.

CI is not a technical thing – it is a cultural thing as well, that promotes shared ownership, encourages collective responsibility for quality and the goal of continuous improvement.

Continuous Deployment (CD): Delivering Value at the Speed of Innovation

CI is about helping code integrate and to be tested, Continuous deployment further automates delivery of validated code changes to production or staging environment. CD makes sure that any successful build from the CI phase is auto released without any manual intervention. This capability makes it possible for organizations to take and to keep users updated faster and more reliably as they do in the present state of development paradigms centered on user effects and radical releases.

Finally, in the depicted workflow, after successful integration, testing and validation, the final validated code is deployed to a production environment. By moving from development to delivery, there are no delays, minimal risk in manual releases, and users get the improvements as soon as they're ready.

Advantages of Continuous Deployment include:

- **Accelerated release cycles**: Software is delivered more frequently, allowing faster feedback and innovation.
- **Consistent and repeatable deployments**: Automation ensures that deployment processes are not dependent on individual developers or operations teams.
- **Lower risk of deployment failure**: Smaller, incremental changes are easier to debug and roll back if necessary.
- **Improved customer satisfaction**: New features, enhancements, and bug fixes reach users more quickly and reliably.

CD changes deployment into a routine, low risked activity rather than a hazardous, high awaited one.

Automation: The Pillar of Reliability

CI/CD pipelines are the primary strength of the software delivery process, as they allow the automation of the process which results in faster and more reliable delivery. Any manual tasks in code compiling, running tests etc and deploying or configuring the artifacts are prone for human errors and slow and time consuming as well as are not comparable. CI/CD gives us predictability, efficiency and traceability to the software development life cycle by automating these steps.

Automation ensures:

- Reduces bugs specific to environment.
- Reduce the chances that the defects would get to users by testing and validating consistently.
- The ability to monitor in real time and get alerted to problems immediately post deployment.

The automation pipeline is a quality gate that will catch issues early and enforce standards across the board. Additionally, it relieves the cognitive load on developers and operations team who are freed from the boring and repetitive work to freeing themselves up on strategic work.

The CI/CD Ecosystem: Tools That Enable Automation

There are plenty of tools to build robust CI/CD pipelines, and they all come with different features that are aimed to address different needs of the development and operations teams. Some of the most popular tools are usually chosen based on the team's preference, the existing infrastructure, scalability requirements and other platforms integration.

Some of the most widely adopted tools in the CI/CD ecosystem include:

- **CI Tools**:
  - **Jenkins:** A highly customizable and extensible automation server with a rich plugin ecosystem.

- **GitHub Actions:** A unified CI/CD solution built right in to the GitHub repository.
- **GitLab CI/CD:** Provides a complete DevOps lifecycle in a single application, from source code to deployment.
- **CircleCI:** Focused on performance and scalability, suitable for cloud-native applications.
- **Travis CI:** Known for its simplicity and integration with GitHub.

- **CD Tools:**
  - ***Spinnaker***: An open-source, multi-cloud continuous delivery platform.
  - ***Argo CD***: Argo CD is a Kubernetes-native continuous delivery tool for declarative Gi tops workflow.
  - ***Octopus Deploy***: A .NET and Windows based based deployment automation server.

- **Infrastructure & Orchestration Tools:**
  - ***Docker***: Docker provides support to continue from development to production.
  - ***Kubernetes***: Automates deployment, scaling, and management of containerized applications.
  - ***Ansible* and *Terraform***: Infrastructure provisioning and configuration management with Ansible and Terraform.

These tools facilitate teams to map out end to end delivery pipelines that are completely automated as well as auditable, scalable and maintainable.

4.4.1 Building Resilient CI/CD Pipelines

In today's software development, resilience is not limited to applications but also to the delivery pipelines that make them alive. Resilient CI/CD pipeline is one that delivers software consistently, predictably and robustly, regardless of possible failures, such as flakiness of the tests, issues with infrastructure, as well as unfortunate misconfigurations of the environments. Such pipelines are known to be characterized by fault tolerance, self-healing capabilities, observability and reproducibility. Often these pipelines will isolate errors and contain them, retry failed steps when needed or when appropriate, and give developers feedback real time, with out stopping the entire pipeline from running.

In order to build resilience in CI/CD, teams have to get into a few strategic practices. Breaking the pipeline into discrete steps helps to ensure that one failure in the pipeline is not fatal. For instance, retry logic for transient errors, containerized environments for consistency, and categorize tests for optimized execution make it stable. Similarly, it is equally important to observe, i.e., tools that monitor pipeline performance, test results and deployment logs to catch issues early and address them fast. In addition, the entire application code, as well as the deployment configuration, is version controlled; debugging and rollback becomes easier when failures occur.

A proactive approach to failure and time improvement is also needed in order to have resilience. Logs and alerts should be informative and use pipelines should fail fast and clearly in order to resolve problems quickly. Build agents are redundant, pipeline infrastructure is high available and artifacts are distributed in storage for redundancy. Reability is strengthened by security, access control and automated rollback mechanisms. Finally, resilient CI/CD are not just technical assets, but also cultural and operational investment to allow a development team to innovate faster whilst retaining system integrity and cultivating trust with the customer.

4.4.2 Canary Releases and Blue-Green Deployments

Advanced deployment strategies are Canary releases and blue-green deployments that prevent the introduction of errors into production environment. These techniques enable the teams to validate new features in a controlled way and guarantee the system stability before the change is fully rolled out. Both methods are largely used to minimize the downtime, increase user impact during update and revert roll back in case of any failure, which is very important for the current Site Reliability Engineering (SRE) and DevOps practices (Winn, 2017).

Canary releases are where you release new version of software to a small user group before releasing it to the entire user base. This enables testing in real world in production without affecting all users. To monitor, performance, error rates, and user behavior, metrics, observability tools are used. The release can be halted or rolled back in case of anomalies with minimal disruption. Specifically, this is a good approach when you want to validate features, improve performance or modify configuration in a way that is safe and confident.

However, blue-green deployments deploy two identical environments that they name "blue" (the current live environment) and "green" (the new version). It is deployed to the green environment and is rigorously tested. After validation, traffic is switched from blue to green and the green environment becomes the new production system. If issues arise, the blue environment is still available as a backup, which can be used to roll back an instant. It keeps the environment simple and releases zero downtime. Both strategies used effectively increase the reliability, agility and resilience of deployment pipelines.

4.4.3 Avoiding Deployment Downtime with Rolling Updates

A deployment strategy of rolling updates refers to the incremental deployment of software updates on servers or instances instead of all at once. This is basically to keep uptime of the application and reduce downtime during the deployment. Instead the system updates a few instances at a time while the rest serve traffic. This staggered rollout approach protects users from a lack of service disruption as a result of the release of a major (Liang, 2021).

A common deployment practice is to manage the rolling update process of the deployment on an orchestration tool like Kubernetes, AWS ECS, or

Docker Swarm in a rolling update. These tools know how to steer traffic away from the instance that is being updated so that it can be replaced with a new version and only proceed with the next batch of instances after they have health checked that they are stable. Teams can halt the rollout of such a huge project at any time if they see any of the system metrics such as latency, error rate, and system load falling. However, this also turns rolling updates into an effective approach to keep a system running reliably in high availability systems, especially in production.

The support for partial rollbacks is another one of the key benefits of the rolling updates. Once the rollout begins, if a failure or bug is found midway through, it is possible to roll back the updated instances without affecting the entire production environment. This decreases risk and increases confidence in continuous delivery practices when combined with good observability and alerting. Rolling updates provide good agility and caution by preventing teams from being too agile while not creating downtime or poor user experience.

4.5 Self-Healing Systems and Auto-Remediation Techniques

In modern systems that do scale and get more complex, it is no longer sufficient to use the traditional reactive approach to incident management. The paradigm shift for self-healing systems is to proactively detect, diagnose and resolve failures which does not require manual intervention. Failure is anticipated to occur as a natural part of distributed computing in these systems, and they are architected to respond to those failures with built in intelligence and automation. The principle of autonomy lies at the heart of a self-healing system that allows the infrastructure and the applications to not only detect anomalies such as degraded service, failed applications and broken dependencies, but also allow them to take corrective actions automatically. The goal is final to guarantee the service continuity and reduce downtime, improving the trust towards users and operation economy (Jackson, 2009).

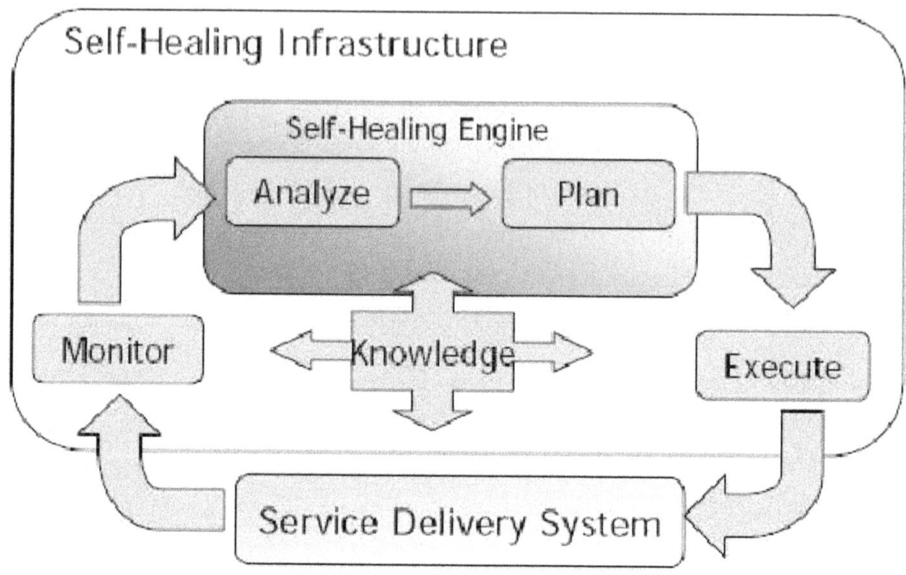

Figure 4.7: Architecture of a Self-Healing Infrastructure System.

Source: (Montani & Anglano, 2008)

However, organizations often integrate auto-remediation techniques on these CI/CD pipelines, orchestration platforms and monitoring systems to be able to implement self-healing capabilities. Monitoring tools such as Prometheus, Datadog, AWS CloudWatch are used to monitor the metrics and log patterns. If a problem is detected (e.g a spike in CPU use, failure of a health check or the inability to reach a particular api) the system has a preconfigured workflow to resolve that problem. To give an example, Kubernetes can automatically restart or reschedule failing pods, an auto-scaling group launches new instances (instances) when pod capacity is lost, or an infrastructure as code tool such as Terraform will do what is configured and re-provision incorrectly configured resources. In the more advanced cases such as AWS Lambda or Azure Functions, these remediation logics, like flushing a queue, purging a cache or resetting a database connection pool can be ran immediately once an issue is detected to prevent data loss. The result is even more resilient and reliable system by reducing Mean Time to Detect (MTTD) and Mean Time to Recovery (MTTR).

To build robust self-healing systems, planning must be done with care, intelligent design must be achieved, and validation must be continuous. A foundation to start with, observability narrows down to emitting rich telemetry data (metrics, logs, traces) and let automation make decisions based on them. But equally important is defining what is an unsafe and unreliable remediation action, perhaps restarting an experiment that failed but following a safe based procedure to container deployment might be too hazardous, and other actions that require some safeguards or a approval workflow, like pulling a container down and restarting a database. Throttling mechanisms and circuit breakers are also necessary to avoid recursive failures (remediation storms). However, various machine learning models and anomaly detection algorithms can be implemented in organizations as systems mature which allows the use of predictive auto-remediation to predict potential failures prior to them occurring. Using these intelligent automation practices, enterprises can reduce the workload for SREs, achieve better stability of the systems, and develop the digital services that recovery gracefully and continuously tune themselves.

4.5.1 The Concept of Self-Healing Infrastructure

Self-healing infrastructure is defined as the systems that are designed with mechanisms to detect, diagnose and repair a fault or failure from performance issues without human intervention. In such a world where services are distributed and interdependent, this concept is becoming ever more crucial. Humans had to wait until issues appeared, and take corrective measures, using what was typically a reactive approach for traditional infrastructure. On the other hand, the self-healing infrastructure actively monitors itself and address failures with real time behavior trying to maintain system availability and reliability despite the irregular failures and Occurrence of errors (Baines et al., 2014).

Real time monitoring, alerting and automation frameworks are the main components of the core of a self-healing infrastructure. The data generated from these systems are constant and it collects and analyzes health signals such as system metrics, application logs, event trace, and user experience indicators. Automated remediation workflows are launched when an anomaly occurs, for example a node going offline, health check can fail a service, or significant increase in latency. It might be doing something as simple as a restart of services, to re-provisioning of the virtual machines, rolling back deployments, to rerouting traffic to healthy nodes. On the other hand, tools like Kubernetes, AWS Auto Scaling, Google Cloud Operations Suite, and Terraform make it possible to integrate the observability with dynamic provisioning and infrastructure as code automation to bring the self-healing capabilities.

In addition, the self-healing model changes the role of the engineer from being constantly on the firefighting mission, to one of architecting intelligent, resilient systems. It promotes a sense of proactive reliability engineering where failure is anticipated and created. It also enables scalability and agility via minimization of bottlenecks during faults and infrastructure up, detracting overhead and improving uptime. As machine learning allows self-healing mechanisms to become more sophisticated, over time, they are able to incorporate machine learning so that they can make better predictions of failures and better choices for remediation strategies.

4.4.2 Using AI for Auto-Remediation

Since systems will be more and more complex and scaling, traditional rule-based auto remediating techniques are not broad enough and are not excessive. This is where Artificial Intelligence (AI) takes over to automate the auto-remediation process to a more dynamic and intelligent one. By applying machine learning (ML), anomaly detection and pattern recognition, AI driven auto remediates the behavior of systems as existing over time. Rather than use fixed sets of rules it can respond to changing conditions, patterns of subtle deviance from normal behavior and intentional trigger remediation actions more precisely and contextuality. This helps to reduce false positives and increase response accuracy in particular in a distributed environments with issues being non-linear and interrelated (Arora et al., 2024).

Continuous learning happens in AI based systems with inputs of historical incidents, logs, etc. to create models to find possible failures because they have occurred in the past. An example of this is an ML model learning that a combination of higher CPU usage and longer database queries lead up to a service outage. In the event of those conditions again, the system can be scaled proactively with resources, services can be restarted, or traffic can be routed in order to avoid an incident altogether. AI is used to correlate events across multiple systems, rank alerts according to business impact, and recommend or take remediation steps in all sorts of platforms like Moogsoft, Big Panda, AIOps capabilities in tools like Dynatrace and Splunk, and so on. This automation does a lot to reduce Mean Time to Detect (MTTD) and Mean Time to Resolve (MTTR) so systems are healthy and performant under stress.

The key power of AI to auto-remediation is its ability to deal with complexity and uncertainty. In large-scale environments of thousands of microservices, it's nearly impossible for human operators to disrupt and act on all the problems in real time. Root causes and corrective actions stemming from the analysis of billions of lines of code across a range of logs and metrics, and can do so much faster than a human could even identify the problem. Furthermore, AI systems can be evolved through reinforcement learning by learning which particular actions give better outcomes and repeating these actions while continuously refining those which gave bad outcomes. This feedback loop creates a situation in which the remediation strategy becomes better over time. Adding AI to auto-remediation brings us closer to self-managed infrastructure that not only provides resilience, scalability, operational efficiency, but is a system characteristic itself.

Multiple Choice Questions (MCQs)

1. **What is the primary benefit of automating repetitive operational tasks in Site Reliability Engineering?**
   A) Increased manual oversight
   B) Reduced system observability
   C) Improved consistency and reduced human error
   D) Elimination of all engineering roles
2. **Which of the following is a potential risk of over-automation in a production environment?**
   A) Reduced deployment speed
   B) Complete elimination of downtime
   C) Loss of visibility and control over processes
   D) Decreased configuration complexity
3. **Which IaC tool is declarative, widely used for provisioning infrastructure, and works with multiple cloud providers?**
   A) Jenkins
   B) Kubernetes
   C) Terraform
   D) Prometheus
4. **What is a best practice when managing Infrastructure as Code (IaC) at scale?**
   A) Avoid using version control
   B) Deploy changes directly in production
   C) Use modules and maintain reusable code
   D) Write infrastructure manually in cloud consoles

5. **Which practice ensures traceability and accountability when working with Infrastructure as Code?**
    A) Dynamic resource allocation
    B) Version control using Git or similar tools
    C) Real-time debugging in production
    D) Manual configuration backups
6. **In configuration management, which tool is best known for its push-based model and agentless architecture?**
    A) Ansible
    B) Puppet
    C) Kubernetes
    D) Jenkins
7. **What is the main advantage of using Kubernetes in configuration management and deployments?**
    A) Limited scalability
    B) Centralized database for logs
    C) Orchestration of containerized workloads for scalability
    D) Manual network provisioning
8. **What is the purpose of blue-green deployments in CI/CD pipelines?**
    A) To update only the staging environment
    B) To switch traffic between two identical environments for zero-downtime deployment
    C) To deploy updates without any monitoring
    D) To delay code reviews until after deployment

9. **Which deployment strategy gradually replaces old application instances with new ones to prevent downtime?**
   A) Immutable deployments
   B) Canary releases
   C) Rolling updates
   D) Shadow testing

10. **How does AI enhance auto-remediation in self-healing systems?**
   A) By replacing all human decision-making
   B) By manually executing test scripts
   C) By predicting failures and initiating corrective actions autonomously
   D) By scheduling weekly maintenance windows

Answers

1.	2.	3.	4.	5.	6.	7.	8.	9.	10.
C	C	C	C	B	A	C	B	C	C

CHAPTER 05

SRE Best Practices and Organizational Adoption

5.1 Implementing Error Budgets and Balancing Reliability with Innovation

The error budget is the unreliability that is allowed in the system without disappointing the customers. It gives a clear value a service can have that is still acceptable in terms of unreliability and thus provides a metric by which teams can measure their system's performance. In a sense, error budgets enable teams to strike a balance between ensuring reliability and the need for speed of innovation, and they provide guidance in where to apply efforts. Error budgets are a strategic tool for system reliability management. However, to a certain extent, they let developers make mistakes and innovate. As soon as the system is depleted the focus turns to increasing system reliability. Such dynamic approach ensures user trust and satisfaction, and implements a culture of continuous improvement by calculating error budget (Penny W. Cloft, Michael N. Kennedy, 2018).

An error budget is a definition of a level of unreliability you can have within a service and still be good to the user. This concept of balancing innovation and reliability means teams can concentrate on work, maintain a reliable service. The service level objective (SLO) of a system is used to calculate error budgets based on the allowable level of failures for a system. Teams that maintain an error budget can make sure their services are in acceptable reliability limits and not be left with user dissatisfaction during outages or failures.

Importance of Error Budget Policies

System reliability and speed of new feature delivery need to be balanced by error budget policies. By establishing a clear error budget policy, organizations can optimize between innovation and service reliability and satisfy customers. These policies define what response to take when the error budget is exceeded, so that corrective actions can be taken and further problems can be avoided.

Alerts when error budget is almost spent can help minimize the impact on the customers. It is helpful in alerting management about an approaching exhausted error budget, which helps to decide the right action to be taken. The proactive approach ensures that the organizations have high service reliability to meet customer satisfaction.

How to calculate an error budget

Define the SLO: The first part of determining an error budget is to define the SLO. It should include a clear definition of what the service is and how much availability or uptime is needed:

1. **Determine the acceptable error rate**: The next step after defining the SLO is to set the acceptable error rate. The maximum rate of errors or downtime allowed within the SLO is this.
2. **Calculate the total allowed error or downtime**: Find the total allowed error or downtime of the SLO using the acceptable error rate. That can be done by multiplying the acceptable error rate by the length of time (e.g. one month).
3. **Measure the actual error or downtime**: Then we measure the actual error or downtime at the SLO during the same time period. There are many means of doing this using different types of monitoring tools and techniques.
4. **Calculate the error budget:** To determine the error budget, remove the actual error or downtime from the total allowed error or downtime. This remaining budget is the amount of error or downtime still acceptable for the SLO.

By following these steps, teams can calculate an error budget for their SLO and use it to manage the reliability of their services. It's important to note that error budgets should be revisited and updated regularly to ensure they remain relevant and effective. Additionally, error budgets should be set realistically and take into account the specific needs and goals of the service or system being managed.

5.1.1 Using Error Budgets to Balance Stability and Speed

Error budgets are a powerful way to find a stability development speed tradeoff (reliability, uptime, performance vs new features, deployments, changes) through a process of elimination. Site Reliability Engineering (SRE) at Google has this central to it, but it is applicable to any organization that's managing software systems (Smith, 2021).

Error budgets act as a strategic tool to balance innovation with demand for reliability of a system. Error budgets allow teams to compute risks to give them a measured amount of non-responsible downtime, which allows innovation without impacting customer satisfaction. The button that represents this balance is key to keep users trusting and simultaneously building a culture of continuous improvement and accountability. Also, error budgets help control risks, schedule work, and enhance team dynamics. This other aligns development and operations teams aiming at common reliability goals, creating a culture of accountability. This approach of collaboration makes both teams focused on the improvement of system reliability and user satisfaction.

Error budgets are a performance indicator and guide for system development efforts. Maintaining balance: They balance between producing new items and the reliability. That allows organizations to see how sustainable the current development pace is. Creating a balance between releasing something new while keeping the system stable for the long haul, the teams can predict and trade off how much unreliability to accept over a specific time frame. Error budgets are a powerful tool for reliability management of high availability systems. By establishing an error budget and well managing it, teams can control innovation with reliability and keep their services intact and available to customers over time. This helps the customers, stakeholders, and the team members to build trust and confidence.

Teams can use error budgets to effectively implement the budgets and to do so, they should have clear SLOs, time-based error budgets, regularly revisiting and updating their error budgets, prioritizing the improvements and fixes, communicate with stakeholders, and use error budgets with other metrics. Implementing these best practices will allow teams to effectively manage the reliability of high-availability systems and make sure that their services meet the requirements of their customers and stakeholders. However, with continued evolution of technology and increasing customer expectations, error budgets will become increasingly crucial in setting the reliability limits of high availability systems. Teams can develop reliable, scalable, and resilient systems that meet their customers and stakeholders' needs over time by adopting error budgets and adopting best practices (Best practices) for error budget management.

A solution to support teams using SLIs, SLOs and error budgets is Harness Security Reliability Management (SRM). It also enables teams to get to the point of applying SLO policies to automate guardrails within CI/CD pipelines. Explore Harness SRM, part of the Harness Software Delivery Platform, along with a demo request today.

5.1.2 Managing Risk While Maintaining Innovation

The caution that organizations need to install in such a culture to achieve excellent results in risk management and innovation. A risk aware culture does not discourage taking chances but rather it fosters a culture that attempts to make informed decisions about action before taking that action by thinking through the what impact might occur from the actions under consideration. This requires telling the teams about the risk identification, encouraging open discussions on uncertainties, and promoting transparency when potential issues occur. For this type of behaviour to be modelled, leaders must emphasize that learning from setbacks is as valuable as celebrating successes. Creating such an environment, where risks are visible and not closed the eyes to, but are managed with foresight and responsibility, is enabled when employees feel safe to report concerns and to share insights (Sutton, 2002).

Risk assessment should be a continuous process that will be done from the beginning of the project development stage to the end of the project cycle. Given, the identification of risks early allows organizations to set up mitigation plans before things escalate. Strategy that can be taken proactively may include scenario planning, contingency budgeting, redundancy in system design, or installing early warning mechanism through observability tools. Also, organizations should regularly review and post mortem other critical incidents afterwards to refine the understanding of risk and make the organization more resilient in the future. These assessments must be revisited in projects since the project has evolved, and new risks must be captured and addressed as they happen.

In the end, effective risk management is about minimizing risk tolerance while minimizing reward and vice versa. When innovation involved such process, stepping into the unknown is inherent and organizations that are too risk averse may miss transformative opportunities. Teaming can handle strategically making the trade-offs between the safety and advancement to foster the continuous innovation. A mature risk management framework allows organizations to make judicious decisions at the risk, grow their capacity to learn from outcomes and to move towards long term goals in a reliable and value enhancing way.

Figure 5.1: Steps of Risk Management.

Source: (Baughan, 2023)

- **Risk Identification:** Understanding what could threaten an organization, its operations and people. It includes holding practices like evaluating IT security threats or observing for events that could interfere with business operations.
- **Risk Analysis:** Identified risks to be examined in order to understand the impact they might have on the business or project. In this step, the risks and their likelihoods and consequences is determined.
- **Risk Prioritization:** Identifying risks and categorizing them by their severity and likelihood to deal with the highest risks first. It guarantees that the most serious threats are well managed using resources effectively.
- **Risk Mitigation:** Strategies to minimize identified risks, developing and implementing them. Strategies can be devised to avoid, reduce, transfer, or to accept the risk.
- **Risk Monitoring:** Staying on top of and reviewing the risks that were identified, and how the mitigation strategies were being

effective. This step ensures that the organization is prepared to respond to changes in the risk profile over time.

Although I never liked taking chances, I now understand that they are necessary for an innovative and creative process. The secret is to comprehend the methods needed to control hazards in order to promote creative fixes for current issues. For complicated Enterprise Partnership initiatives, wishing for the best is not a risk management approach. These are my top five suggestions for controlling innovation risk.

5.2 Collaboration Between Development and Operations Teams

Collaboration between the development (Dev) and operations (Ops) teams is essential to providing high-quality products and services in Site Reliability Engineering (SRE). Together, these two historically disparate teams have created a ground-breaking plan for advancing a culture of automation, cooperation, and constant development. In old IT setups, development teams were primarily concerned with pushing features, while operations teams were in charge of stability and uptime. This separation often led to misaligned goals, delayed releases, and more finger-pointing when things went wrong (Niall Richard Murphy, Chris Jones, 2016b).

This partnership makes handling questions and comments quick and efficient. While operations teams receive early notice of impending changes, development teams obtain a greater understanding of how their code performs in production. Reducing mean time to detect (MTTD) and mean time to remedy (MTTR) problems is essential for strengthening systems and improving user experiences.

Additionally, as the majority of organizations are using various incident management, monitoring, and observability platforms, downtime expenses are reduced. By integrating these platforms with automated warning systems and CI/CD pipelines, both teams can react to abnormalities instantly, avoiding extended outages and guaranteeing that service level objectives (SLOs) are regularly fulfilled.

5.2.1 Bridging the Gap Between Dev and Ops

DevOps is a collection of practices and philosophies to bring software development (Dev) and IT operations (Ops) together as a single, singular team. DevOps doesn't work in silos, it advocates for cross functional collaboration and communication, the whole application lifecycle is shared sense of responsibility. This holistic approach simplifies the process, reducing the time and reliability of high quality software delivery for organizations (S. Sharma, 2017).

Key Principles of DevOps Followed at Immersant

- **Collaboration:** Developers and ops team can work together in all stages of software development lifecycle in DevOps. This collaboration makes it easier to deploy and maintain the software.
- **Automation:** Automation is one of the core principles DevOps. It entails automating the repetitive task like building, deploying code, testing and provisioning infrastructure. Automation decreases errors, increases speed, enhances the overall efficiency.
- **Continuous Integration and Continuous Deployment (CI/CD):** DevOps is filled with CI/CD. Continuous Integration means integrating code changes automatically into shared repository and running the automated test. Continuous Deployment goes one step further than this: it automatically deploys changes in code to production environments as soon as it passes the tests.
- **Monitoring and Feedback:** Monitoring tools are used by the DevOps teams to discover real time information about about the current application performance and infrastructure health. It is a kind of feedback loop through which teams can predict and deal with problems to improve system reliability.

The four phases of DevOps

The process through which DevOps evolved has shown four phases; the first being the technology change; the second, the organizational change; third, the integration of both technology and organization change; and finally, the adoption change. The progression of this is a result of increasing devops complexity which arises from two key trends:

1. **Transition to Microservices:** Due to the switch in organization from their monolithic architecture to a more flexible microservice past, there has been a strong demand for the specialized DevOps tools. The goal of this shift is to accommodate the higher grain and agility facilitated by microservices.
2. **Increase in Tool Integration:** There has been a large amount of integrations of Projects and tools as a result of the proliferation of projects and the need for more DevOps tools. This complexity has motivated the organizations to view their approach to DevOps tools as not flexible.

There have been four phases of evolution of DevOps, each solving the increasing needs and complexities of software development and delivery.

The four phases are as follows:

- **Phase 1: Bring Your Own DevOps (BYOD)**

During Bring Your Own Devops each of the teams selected the tool of their choice. However, teams found it hard to work together because they were not used to the tools of other teams with this approach. In this phase, the need for a more unified tool set has been realized, so as to assist with smoother team integration and project management.

- **Phase 2: Best-in-class DevOps**

To address the challenges of using disparate tools, organizations moved to the second phase, Best-in-class DevOps. In this phase, organizations standardized on the same set of tools, with one preferred tool for each stage of the DevOps lifecycle. It helped teams collaborate with one another, but the problem then became moving software changes through the tools for each stage.

- **Phase 3: Do-it-yourself (DIY) DevOps**

To remedy this problem, organizations adopted do-it-yourself (DIY) DevOps, building on top of and between their tools. They performed a lot of custom work to integrate their DevOps point solutions together. However, since these tools were developed independently without integration in mind, they never fit quite right. For many organizations, maintaining DIY DevOps was a significant effort and resulted in higher costs, with engineers maintaining tooling integration rather than working on their core software product.

- **Phase 4: DevOps Platform**

A single-application platform approach improves the team experience and business efficiency. A DevOps platform replaces DIY DevOps, allowing visibility throughout and control over all stages of the DevOps lifecycle.

By empowering all teams – Development, Operations, IT, Security, and Business – to collaboratively plan, build, secure, and deploy software across an end-to-end unified system, a DevOps platform represents a fundamental step-change in realizing the full potential of DevOps.

GitLab's DevOps platform is a single application powered by a cohesive user interface, agnostic of self-managed or SaaS deployment. It is built on a single codebase with a unified data store, that allows organizations to resolve the inefficiencies and vulnerabilities of an unreliable DIY toolchain.

5.2.2 Effective Communication Strategies for SRE Team

Effective communication is critical for the success of Site Reliability Engineering (SRE) teams, given their role in maintaining the reliability, scalability, and performance of services. Poor communication can lead to misunderstandings, delays in incident response, and inefficient collaboration across teams. On the other hand, clear, timely, and purposeful communication helps SRE teams work cohesively, align with organizational goals, and continuously improve service reliability (S. Sharma, 2017).

The first is that Site Reliability Engineering (SRE) covers a wide spectrum of responsibilities and approaches. The functions of SRE are distributed across infrastructural teams, service-oriented teams and horizontal product teams. The working style varies also in the collaborations with product development teams, from working with much larger counterparts to partnering with teams of similar size, and even working as the main product development team. The SRE teams usually have team members with expertise in, systems engineering, software development, architectural design, project management and leadership. Among them, many also have valuable experience in a broad spectrum of industries. Unlike a single organizational model, SRE configures adaptively to the needs based on the fundamentally pragmatic and flexible philosophy.

SRE also doesn't act like being a command-and-control organization. SRE teams generally work in a dual allegiance, service or infrastructure SRE teams work very closely with the product development teams who own the services or platforms that they are responsible for, but they are also part of the SRE organization and its framework and values. SREs are truly accountable for reliability and performance of those systems, and the service level relationship is especially robust. However, their formal reporting lines are still within SRE and centralized. While they spend most of their time nowadays working on particular services, a culture and a set of common values throughout the SRE organization produce the same consistent and effective ways of solving problems.

The two preceding facts have affected SRE organization in two essential aspects of team operations, communication and collaboration. Data flow is a good computing metaphor for communication in SRE teams: data must flow through production systems, and information should flow through the SRE team's projects, service states, production environments, and individual status. The data in this must flow reliably between all relevant stakeholders in order to ensure maximum team effectiveness. A useful way to think of this flow is to think of the API that an SRE team presents to other teams. The same goes as with an API, effective design is important for easy enablement, and a poor design to be fixed later can be costly and difficult.

Collaboration also applies to the API-as-contract metaphor, and also between SRE and product development teams where all parties must move in an environment of constant change. In this sense, SRE team collaboration is also similar to the collaboration in other fast paced companies. Yet the difference is in the particularly good mix of software engineering knowledge, systems engineering understanding, and practical experience in production from SRE. Addressing the production concerns and product goals in a mutual respect environment yields the best designs and implementations. The promise of SRE is that with the same skills as product development teams, an organization responsible for reliability should lead to measurable improvements. The experience tells us that assigning someone the responsibility of reliability is not enough if you don't give them the full skill set.

5.2.3 Encouraging a Shared Responsibility Model

The shared responsibility model is a framework that specifies who is in charge of protecting certain elements of the cloud computing environment—the client or the cloud service provider (CSP). The customer is responsible for protecting its cloud-hosted data and apps, while the CSP is often in charge of protecting the underlying infrastructure. (III, 2018).

An agreement between a cloud service provider (CSP) and an end-user of its services is known as the shared responsibility model (SRM). According to this agreement, an end-user is in charge of protecting the workloads that are executed on the cloud platform, while a CSP will be in charge of protecting the platform infrastructure of its cloud operations.

In fact, Gartner emphasises how important it is for CSP clients to fully comprehend the terms of the contract, pointing out that CSPs are unable to provide total security and that security executives need to be aware of the extent of their duties regarding cloud security. This is particularly true for a company that is moving all or some of its workloads to the cloud.

As a result, it would be ideal for cloud architects to consider the particular security implications of the environment they wish to work in. This will make it easier for all parties involved to understand the risks and obligations the company is taking on when it migrates to the cloud. Insufficient comprehension of the SRM concept in relation to a particular organization and its CSP may lead to the misunderstanding that the CSP is in charge of a particular area's security, which may result in incorrect settings and/or inadequately secured cloud assets.

Familiarizing oneself with one's role in the Shared Responsibility Model (SRM) will assist in ensuring that responsibilities towards the Cloud Service Provider (CSP) are fulfilled and cloud security best practice, such as regular vulnerability scanning, are taken and enforced.

Benefits of the Shared Responsibility Model

- **Scalability:** Customers can increase or decrease security capabilities inside a CSP ecosystem's platform according to their needs at any given time. Large cloud providers are naturally able to grow to meet the changing operational needs of an organization. The security architecture of a CSP will always be established in accordance with the SRM. Customers can therefore expand their own data security policies as necessary with confidence.
- **Collaboration:** In terms of security, the SRM promotes clarity in responsibilities, as previously said. Therefore, such division benefits a customer's business. Compared to when a customer constructs on-premise operational capabilities from scratch, the cloud operations SRM has a lower security load.

- **Architectural strength**: Customers who are concerned about the security of their data on a provider's cloud should feel much more at ease when they work with a reputable and trustworthy CSP. One great advantage of the SRM is that it allows you to fully utilise the security data analytics tools and robust architecture of a CSP.

Shared Responsibility Model in Practice

A more technical summary of what the SRM usually includes would be that, according to many experts, the client is in charge of everything they may add, remove, modify, or reorganize in their cloud environment. It is likely that the CSP will be responsible for overseeing that component of cloud operations if they are unable to make any changes.

However, as previously said, there may be some areas of overlap. For operations to go as smoothly as possible, both CSPs and their clients must have a thorough understanding of these grey regions, also referred to as shared control areas. For instance, patch management, infrastructure as code (IaC) configuration management, and security awareness training are examples of shared control domains in relation to AWS.

In particular, clients are in charge of patching their guest operating system and apps, while AWS would be in charge of patching and repairing defects in their infrastructure. In a similar vein, AWS keeps its infrastructure configured, but each customer is in charge of setting up their own databases, operating systems, and apps.

Finally, it is the responsibility of both AWS and its cloud clients to train their respective employee organisations on security awareness. The ability of CSPs and their clients to secure the areas for which they are exclusively accountable is only strengthened by these shared control areas.

5.3 Building a Culture of Reliability and Resilience

The culture that surrounds such tools and architectures in any complex system—be it technological, organizational, or industrial—is the foundation of long-term success. A culture of reliability and resilience is built by principles, values and practices of consistent performance, proactive risk management, and adaptive response to change or failure (Lee, 2016).

According to many experts, the client is in command of whatever they may add, remove, alter, or reorganize in their cloud environment. This is a more technical overview of what the SRM typically contains. If they are unable to make any modifications, the CSP will probably be in charge of managing that aspect of cloud operations.

There might be some areas of overlap, though, as was already mentioned. Both CSPs and their clients need to have a solid grasp of these grey areas, also known as shared control areas, in order for operations to run as smoothly as possible. Examples of shared control domains with regard to AWS include patch management, infrastructure as code (IaC) configuration management, and security awareness training.

Specifically, AWS would be responsible for patching and fixing issues in their infrastructure, while clients would be responsible for patching their guest operating system and apps. Similarly, AWS maintains its infrastructure, but each customer is responsible for configuring their own operating systems, databases, and applications.

Lastly, it is the duty of both AWS and its cloud customers to provide security awareness training to their respective staff groups. These shared control regions only enhance the capacity of CSPs and their clients to protect the areas for which they have sole responsibility.

One might question the need for an entire culture cantered around reliability, suggesting that adding more automated test cases or integrating tools into the CI/CD pipeline could suffice. However, reliability extends far beyond simple test coverage or automation. It is influenced at every stage of the software development lifecycle (SDLC), beginning with system design and continuing through implementation, testing, deployment, and operations. Issues related to reliability, including defects and failure modes, become increasingly costly and complex to resolve when discovered later in the SDLC—particularly if they manifest as production incidents. Therefore, fostering a culture of reliability ensures that potential risks are addressed proactively and systematically, rather than reactively.

Second, modern applications and systems are more complex and have more interconnected parts. While traditional testing is good at testing individual components, it's inadequate at testing an entire system holistically. Improving reliability means testing and strengthening these complex interactions to prevent failures in one component from bringing down the entire system.

Lastly, organizations tend to prioritize other initiatives instead of reliability priori, such as shortening development cycles and quickly releasing new features. This isn't because reliability isn't important, but rather that it isn't a top priority for many teams. Without a strong incentive coming from the organization, efforts and initiatives to improve reliability are less likely to maintain momentum. Faster feature development can even hinder reliability efforts by making systems more unreliable.

Steps to Build a Reliability Culture

In order to build, foster, and sustain a true reliability culture, actions speak louder than words. Initiatives that are so far-reaching must start at the top, with leadership commitment, and penetrate to every level of the organization. Establishing a reliability culture is not a one-time project; it's a completely new mindset that will need to be nurtured and improved over time for the long run.

Here are some key steps needed for success:

- **Start at the Top with Leadership Commitment**: At the heart of a reliability culture lies a genuine leadership commitment. Leaders set the tone for the organization, influencing the mindset and behaviours of the entire workforce. When leaders prioritize and champion reliability, it becomes ingrained in the organizational DNA.

- **Identify Champions**: Whenever a new initiative is launched, it is common to find enthusiastic team members who grasp and support the vision from the outset. They can be crucial in the momentum by being champions of building a culture of reliability. It helps to fuel early successes and has a spillover effect to expand the organization's engagement and adoption.
- **Define Reliability Expectations:** At the outset, define reliability standards and expectations for all aspects of the operations so that employees have something to fall back on. Then, keep in mind to write down the standardized processes so that there is consistency in work procedures.
- **Communicate Concisely, Constantly, and Completely**: It is important to have effective communication in order to build a reliability culture. The feedback mechanisms are clear expectations and goals that everyone in the organization knows the importance of reliability and the role they play in maintaining it. Encourages open and transparent communication channels all over the business.
- **Share Successes:** There is a lot of work being done to share communication and an effort to share success stories and efforts across the organization. If we have people in the team who are dedicated to reliability and perceived to be getting recognised for it, then the team are more likely to jump on the bandwagon of reliability.
- **Foster Ongoing Learning and Continuous Improvement**: A reliability culture embraces a proactive approach to learning and improvement. It encourages employees to identify areas for enhancement, learn from both successes and failures, and implement changes to enhance overall reliability. Establish mechanisms for receiving feedback on processes and performance,

and embrace an iterative approach to improvement, making incremental changes based on lessons learned and feedback.

- **Identify Risk Management Issues**: The main element of reliability culture is recognizing and managing who risks. Any organization should have sufficient robust risk assessment processes in place in order to assess the potential threats to reliability and devise appropriate strategies to avoid the threats.
- **Empower Employees**: Reliability culture authorizes every employee to contribute to the organization's reliability goals. Empowerment allows employees to feel ownership and accountable, so they are all involved with the problem solving and decision making process. Lay down mechanisms by which employees can give their feedback on processes and indicate the ways to improve.
- **Encourage Data-Driven Decision-Making**: Reliable operations are often data-driven. A reliability culture is one that makes the collection and analysis of relevant data a normal part of the process for decision making, so that actions are data and practice driven.

Benefits of Reliability Culture

- **Enhanced Productivity**: A reliability culture reduces the time of downtime, errors and rework, thus reducing productivity loss. If employees are committed to delivering reliable results, efficiency comes easily.
- **Improved Reputation:** The more an organization consistently delivers reliable products or services, the more its reputation in the market is enhanced. Trust between customers and the stakeholders develop and result into long term relationship and positive brand recognition.
- **Risk Mitigation:** By nature, a reliability culture deals with and minimizes risks, thereby diminishing the chances of costly errors or failures. Such an approach is proactive and it helps organizations tackle challenges easily.

- **Adaptability**: Reliability culture organizations are more likely to be flexible to environment changes. Continuous Improvement and Learning focus allows them to remain on top of things, and be ahead of the curve in fast changing environment.

For a culture of reliability and resilience, leadership must be intentional, the teams must be empowered and there must be an ongoing improvement. When reliability is in the mind of an organization, reliability, trust and long-term success come. In the end, it gives organizations the power to run and repeat, and to pivot fast and alive.

5.3.1 Psychological Safety and Learning from Failures

Learning from failures depends on psychological safety, the belief that one can speak freely and not to suffer the consequences. When there is psychological safety, mistakes can be considered as opportunities for improving our purposes. A culture such as this empowers people to be able to take risk, to voice concerns, and to admit mistakes without being embarrassed or punished. By encouraging this open communication, everyone gains trust, works together, and continuously learns those things that are the big difference between a high performing team and not a high performing team. It is when people are not afraid to fail that they are more likely to innovate, to share unconventional ideas, to learn from setbacks and so create more resilient and adaptive organizations. Additionally, leaders who lead vulnerably, who model reflection, promote psychological safety that makes the culture of blame give away to one of shared accountability and problem that is constructive.

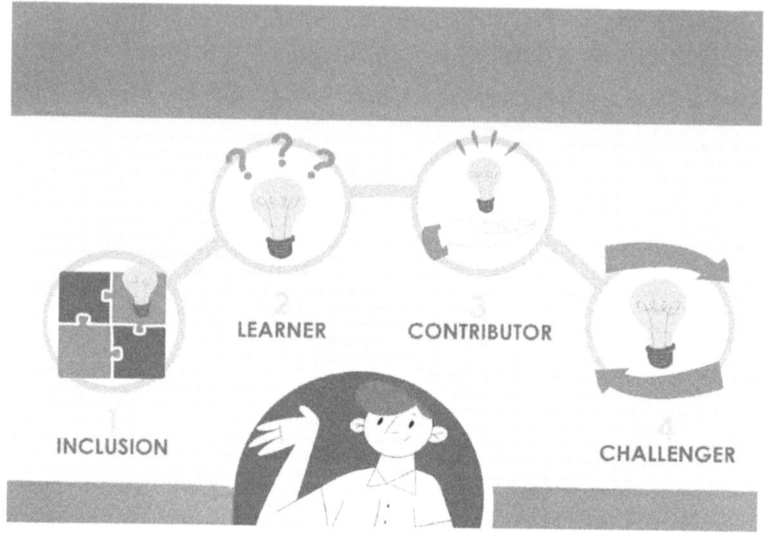

Figure 5.2: Stages of psychological safety.

Source: (Eagle, 2023)

- **Stage 1: Inclusion Safety:**

In this stage, they are looking to feel accepted and to belong to the group. Puzzle pieces fitting together symbolizes that it is important to be included and connected with others.

- **Stage 2: Learner Safety:**

The emphasis of this preoccupation is the capacity of asking questions, experimenting and making mistakes without passing judgment or being equal to punishment. It is represented by the lightbulb with question marks, as it represents the process of curiosity and learning.

- **Stage 3: Contributor Safety:**

In this stage individuals have the sense of safety and are therefore able to contribute their ideas, skills and contributions in the group. A lit lightbulb in the hand signifies that one can share one's talents and insights without fear.

- **Stage 4: Challenger Safety:**

The last of these stages is one in which we feel secure enough to call into question the status quo, comment positively upon areas for development, and urge change. The two arrows pointing in opposite directions indicate the power of questioning and going where no one has gone before.

The stages are arranged in a circular format, which implies that they are interrelated and are built upon each other. The image provides a visual representation of the journey towards creating a psychologically safe environment where individuals can thrive and reach their full potential.

5.3.2 Encouraging Proactive Reliability Improvements

Today, waiting for failure before acting is not an option in fast and complex operational environments. To truly reach the level of operational excellence, business has to stop being reactive and start being proactive, shifting from reactive problem solving to proactive reliability improvement. It implies that foresight, preparedness and continuous innovation become an integral part of business practices. For the purposes of proactive reliability, proactive relating isn't only about preventing failure, but rather about creating systems and teams who are anticipating issues, addressing root causes first, and adapting continuously to meet demands of the future (Rohrbeck, 2010).

Historically, a lot of organizations have been fixing what's broken and restoring systems back to their previous state after the issue has occurred. Although it is necessary, such a reactive approach frequently results in unnecessary incidents, amounts of unnecessary expenditure, and degraded efficiency. Proactive reliability is concerned with potential risks and weak points prior to failure. The strategies include forward thinking practices like condition-based monitoring, predictive analytics, early warning systems to give insights into system health and operation trends.

The culture or an organization must foster proactive reliability improvements, such as foresight, curiosity and continuous learning. It should be trained to employees at every level of the company to be able to identify at an early stage degradation, inefficiencies, or procedural errors. To instill this change in teams, they start asking questions like "What could go wrong?" or "How can this process be allowed to be more robust?" Enabling people to become responsible for reliability not only attracts better outcomes but also creates a better morale. The more autonomous and supported the people feel they have to help them suggest ways to improve the organization, the more engaged and committed they will be to making the organization succeed. The reason managers and leaders should actively solicit input from frontline workers is that they are usually the ones who have hands on experience about the potential risks and inefficiencies.

The ability to proactively make reliability improvements depends to a great extent on technology. Organizations are using modern tools like predictive maintenance system, AI diagnostic systems and real time monitoring platform to detect the anomalies before they become serious problem. In the context of using sensor, operational or performance analytics data inputs, organizations can unearth hidden trends and early warning signs of the potential for instance failures. This proactive approach also reduces the amount of downtime needed in order to rectify an issue that could have been resolved proactively if they were caught sooner. Early detection of such signals in an instance's lifecycle is critical to maintaining high service quality and operational efficiency. Vibration analysis, thermal imaging and lubrication monitoring can for example reveal wear and degradation of equipment before a breakdown happens, in industrial applications. Automated observability tools exist in software and/or IT environments and can find performance bottlenecks, security vulnerabilities and even code regressions as early as possible in the development cycle. However, acting on these insights in a timely manner strengthens the system stability, reduces downtime as well as improves customer satisfaction.

While it is an initiative on a pro-active basis, it is not a onetime thing to do; it has to be a part of growing and maturing as an excellence organization.

What it means is creating structured means for learning from lessons, performing regular audits and improving processes based on feedback. What did not happen should not be the only thing post accident reviews focus on – understand what signals went uncaptured, and what could be done in the future to prevent a similar failure. Organizations also need to establish measurable goals and key performance indicators (KPIs) related to reliability improvements. Mean time between failures (MTBF), first time fix rate, system uptime are obvious metrics that add concrete evidence of progress and can also support the high momentum. Putting these goals into employee evaluations, team goals, and the strategic planning process ensures that proactive reliability continues to be a priority in all functions.

Proactive reliability improvement can only be encouraged by the leadership commitment. The lead must come from the executives and managers, using examples of reliability in decision making, allocating resources for improvement initiatives, and celebrating progress. Rather than only being an optional task, it becomes a strategic imperative when leaders show that proactive reliability is a key goal that teams should strive to fulfill long-term. It is further important for leaders to appreciate and reward teams or people that provide insights about risks early, propose process improvements, or prevent disturbances. This also ensures that desired behaviors are reinforced while at the same time embossing reliability as a central value of the organization.

Finally, the proactive reliability improvement is about promoting the mindset of anticipation, early intervention, and continuous improvement. It gives employees the ability to use technology to instill reliability as a proactive advantage that contributes to long term success, rather than a rigid business necessity to do something in response to the issues created by unreliable behavior. The philosophy of this will help organizations to minimize disruptions and save costs but more importantly will provide the resilient foundation for continued growth and innovation.

Establishing an SRE Mindset Across the Organization

An SRE mindset revolves around the idea that reliability is a shared responsibility—not just the concern of a specific team or a reactive operations function. In traditional IT models, development and operations often operate in silos, with developers focused on delivering features and operations teams responsible for keeping systems running. The SRE approach breaks down these silos, promoting a culture where everyone—from product managers to engineers—takes ownership of the system's health and stability (Blank-Edelman, 2024).

SRE (Site Reliability Engineering) blends software engineering practices with IT operations to build and maintain systems that are not only scalable and performant but also resilient and maintainable. It is using the code to manage infrastructure, doing automation for routine tasks, and engineering rigor for operational difficulties.

It is a mindset to consider reliability early in the system design, development and deployment, not an afterthought. It has a doctrine of measurable service-level objectives (SLOs) and service level indicators (SLIs) to measure performance and error budgets to balance innovation and risk. These concepts allow teams to make the right decisions of reliability over speed versus speed over reliability.

Also, SRE mindset fosters a blameless culture that considers failures as learning opportunities and not as reasons for pointing fingers. After an accident, post incident reviews are used to find systemic faults as opposed to the individual errors, encouraging a continual improvement and more trust between the teams.

To put it more bluntly, the SRE mindset changes the organization's relationship with reliability from reactive firefighting to proactive engineering where the whole organization is driven to build robust, self-healing systems that enable long term growth and customer satisfaction.

Figure 5.3: SRE mindset

Source: (Cortex, 2022)

The image shows the SRE (Site Reliability Engineering) mind set, and these three principles:

- **Encourage ownership:**

This is represented by a man with a trophy who represents the importance of taking responsibility for systems' reliability and performance. It is emphatic in the sense that people and teams should own and improve the services they are managing.

- **Build for scale and invest in automation:**

A robotic arm working on a complex mechanism is depicted as an attempt to design systems that can handle increasing workloads and to automate repetitive tasks. Managing large scale systems is best done with automation to cut toil.

- **Be proactive and be prepared:**

A set of tools and metrics, a timer, calendar, and checklist now illustrate this principle, monitoring systems, anticipating problems, and having plans to deal with them. Desktop tool monitoring and incident management are essential to keeping systems reliable and downtime down.

The three principles that make up the SRE mindset are based on the use of software engineering practices to manage IT operations and deliver reliable services. SRE aims to balance the need for innovation with the need for stability, allowing organizations to deliver reliable and scalable systems.

5.4 The Role of AI and Machine Learning in SRE

AI powered solution brings transformative ability to SRE teams that is beyond being restrict to just monitoring and incident response. While these technologies are based on the power of the machine learning algorithms, they use them to analyze the vast quantities of operational data in real time for detecting anomalies and potential system failures early. AI helps teams figure out patterns that are invisible to human eyes so that they can prevent issues from getting out of hand, thus leading to a reduction of downtime and a higher service availability (Little, 2020).

At today's Site Reliability Engineering, automation powered by AI is very popular. It allows human engineers to focus on the strategic tasks rather than having to respond to the known failure scenarios and therefore faster incident resolution by triggering predefined responses. Dynamic resource allocation of computing resources can also be taken care by intelligent automation systems based on flow patterns to achieve optimized performance and lower costs on infrastructure. They learn from past incidents over time and become better at responding without human intervention to better support a self-healing operational environment.

However, as AI and ML technologies grow only more advanced, autonomous, adaptive infrastructure is the future. Thinking on evolution, organizations will be able to transition towards predictive and even preceptive models where not only they anticipate problems, but also propose or act accordingly. This is a critical enabler for establishing the highly resilient, scalable, and efficient digital services in the competitive environment.

The remainder of the chapter explores some of the key applications and benefits of using AI in Site Reliability Engineering.

1. Predictive Insights: Moving from Reactive to Proactive

Traditional system monitoring tools tend to alert teams *after* an incident occurs, but AI and ML are changing that paradigm by shifting operations to a predictive model. AI-powered tools can sift through massive amounts of historical and real-time data to identify patterns and anomalies that signal potential failures long before they become critical.

However, for instance, it is possible through machine learning to detect small, usually insignificant changes to the server response times or network latencies, giving SRE teams early signals to fix that before they impact the users. This transition from a reactive to a proactive maintenance model significantly cuts down the downtime and brings down a seamless user experience.

2. Intelligent Automation: Speeding Up Root Cause Analysis

Restoring service quickly depends on finding the root cause when a system failure occurs. This is traditionally done as manual effort, that is, squats into digging through logs to understand metrics and correlating events. This process becomes very automated with AI/ML.

AI algorithms are best at finding complex patterns and relationships in enormous data, thus enabling them to find the root cause of incidents by correlating the logs, traces, and metrics. The rapid analysis effectively decreases the Time to Resolution (MTTR) very dramatically, helping teams to solve it faster and retain it better.

Engineers can spend their time working on long term improvements rather than manually searching hours through data to uncover the issue with the use of AI driven diagnostics leading them right to the problem.

3. Capacity Planning and Optimization: Smarter Resource Management

Even a system failure requires that the root cause be found quickly to resume service. This is traditionally done manually—logging, metrics viewing and event correlation. This process becomes very automated with AI/ML.

AI algorithms are best at finding complex patterns and relationships in enormous data, thus enabling them to find the root cause of incidents by correlating the logs, traces, and metrics. It allows teams to reduce the time to resolution (MTTR) significantly faster and with even greater level of precision, thus drastically reducing time to resolution (MTTR).

Instead of running hours of manual search to find problems, the engineers can get AI driven diagnostics supporting them to point to the problem immediately so that they can help to free them from working on long term improvements.

4. Reducing Alert Fatigue: Context-Aware Alerting

The biggest challenge that SRE teams face is alert fatigue whereby you receive so many alerts most of them are low priority or false positives. Recently, alert systems have been constructed using AI/ML that is filtering out noise and focusing on the most important issues that teams need to pay attention to.

These systems are using advanced algorithms which by using historical data and understanding context of different incidents, warn the teams only when requisite. Through intelligent alerting, this reduces noise, allowing SREs to do less work, and in the process, deal with less noise, decreases stress, improves operational efficiency.

5. Self-Healing Systems: Toward Autonomous Operations

It is possible for a system to detect it, diagnose it, and solve it, without human intervention. Self-healing systems are becoming a reality with the help of AI. That is, systems can autonomously identify anomalies, trigger corrective actions such as restarting services or reassigning resources and even applying patches to mitigate further risks with the aid of ML models.

Take for instance, if AI realizes that CPU is overloaded or has a memory leak, it will automatically start taking remedial actions such as provision CPU resources or restart the affected services. This self-healing enables SRE teams to spend time at strategic activities, instead of being firefighting day to day operational issues.

The Future of SRE: Powered by AI and ML

It is obvious that now, Artificial Intelligence (AI) and Machine Learning (ML) are not just upgrading the existing practices, but are actually recasting the fundamentals of how reliability is done. Predictive analytics, self-healing systems, all of these intelligent technologies are giving SRE teams amazing opportunities to keep system reliability, scalability and performance to unparalleled levels.

SREs are now able to anticipate incidents before they disrupt services through those capabilities. AI is able to analyze patterns in telemetry data to predict potential failures and what can or may start to perform preventive actions. Through such shift from reactive to predictive, management is now able to reduce downtime and cast a stronger user trust.

In addition, self-healing infrastructure, which was previously a science fiction, is now a reality. Automated recovery actions like rerouting traffic, restarting service, provisioning more resources are being learned by the ML models to be taken when performance anomalies are noticed. By implementing these self-managing systems, MTTR is brought down drastically and systems operate without human intervention.

It is clear that Artificial Intelligence (AI) and Machine Learning (ML) are not only enhancing the traditional SRE practices but they are actually redefining how the reliability is achieved. Today these intelligent technologies have empowered SRE teams to keep systems highly reliable, quickly and cost effectively scalable and highly performing in the ever-increasing technology complexity realm.

SREs have the predictive ability to predict the incidents before they impact services. AI can predict failure by analyzing patterns in the telemetry data and suggest or even perform preventive actions. Its shift to reactive — rather than predictive — management lowers the downtime, and boosts user trust.

In addition, self-healing infrastructure, a futuristic idea, is becoming real. Automated recovery actions, with their associated workflow and time, such as rerouting traffic, restarting services, and allocating additional resources when performance anomaly is detected are being trained to be taken directly by the ML model. These systems are so self-managing that MTTR is drastically reduced and systems remain running without human intervention.

Furthermore, it is changing observability. With modern applications ejecting massive amounts of data, it is no longer possible for human engineers to process everything in live time. AI takes out noise, relates events and allows SRE teams to make better and faster decisions. The observability is enhanced, and therefore, there is more confidence in production environments and faster incident response.

At the same time, AI also speeds up CI/CD by automating testing, warning risky deployments, and restricting production only to high quality code. Together with the ML driven capacity planning, it makes the resources use efficiently, less cost, and more performance.

Future will bring convergence between SRE and AI/ML which will make it possible for autonomous reliability engineering – systems not only able to detect and fix problems without any help from human beings, but to keep on learning and improving themselves. SREs will be able to shift their attention from managing systems to creating intelligent frameworks which will lead to continuous improvement.

5.4.1 AI-Driven Predictive Maintenance

Predicative maintenance given by Artificial Intelligence (AI) and Machine Learning (ML) is an advance method to safeguard the system by sticking to the Artificial Intelligence (AI) and Machine Learning (ML) process that keeps a check on the system for anomalous behavior forecasting the risks to happen before the actual system damage happens. This approach lets organizations know ahead of time when things are going to need maintenance so that costs are not incurred or maintenance downtime doesn't occur (Lee, 2016).

These AI-driven models continuously learn by spending time analyzing data from sensors, logs, performance metrics and from the historical records the answers hidden there of the early warning signs of the wear, degradation and any bent performance dip. With these intelligent systems, these are the types of changes to behaviour that they can identify: slight temperature increases, abnormal spikes in CPU, memory leaks, or decreasing throughput that a regular monitoring tool would not be able to see. It enables the catching of issues in their infancy which greatly improves system reliability and operational efficiency.

Apart from that, AI powered predictive maintenance helps in better allocation of resources. Due to actual need, these maintenance activities can be prioritized over fixed schedule and thus reduce unnecessary inspections and reduce disruption to ongoing operations. The transition from time based to condition-based maintenance not only helps in saving labour and infrastructure costs, but also enables extension in the lifecycle of the critical components.

Predictive maintenance reduces this dependency because it is a general capability which can be integrated in complex, distributed environments such as cloud infrastructures or large-scale industrial systems. AI can be used combined with automated tools to even initiate corrective actions, e.g., restart service, initiate backups, expand infrastructure, creating a self-healing ecosystem with no human involvement.

As DevOps and SRE practices are being adopted widely within organizations, the predictive maintenance with the help of AI becomes one of the key sides of reliability engineering that enable uptime guarantees, service continuity and make guarantees better user experience. Eventually, it leads to the growth of a culture of proactive problem solving rather than reactive reaction firefighting, which leads to the digital transformation and long-term business resilience.

With the future of evolution of AI and ML models to become even more accurate, accessible and autonomous, the predictive maintenance will redefine systems' management and maintenance across industries.

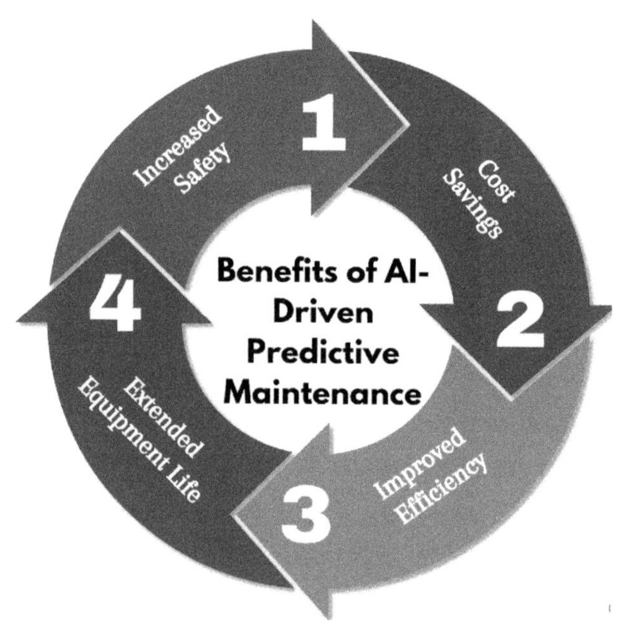

Figure 5.4: Benefits of AI driven predictive maintenance

Source: (Penguin, 2024)

1. Increased Safety:

Predictive maintenance driven by Artificial Intelligence is just one of the most critical benefits as it helps you to improve workers' safety in the workplace. AI systems help to spot early signs of mechanical wear, overheating, pressure imbalances and more before such equipment failures can occur, and lead to accidents, injuries, or even fatalities. This proactive capability of this detection in high-risk environments like manufacturing plants, oil rigs or aviation, functions like an early warning system by enabling teams to intervene prior to any conditions exacerbated into dangerously erratic. Aiding in this is protection of the employees but also compliance with safety regulations and standards.

2. Cost Savings:

Engineered equipment breakdowns can cost millions in terms of emergency repair costs, unplanned loss, revenue loss, and even reputation loss. The risks associated with these failures are reduced by AI driven predictive maintenance that forecasts failure points to allow the organization to schedule interventions strategically. When it needs to be done it is done, without the need for vast amounts of spare parts, overtime labour or costly last minute service calls. As time goes by, this approach results in extreme savings in maintenance related expenses that is the main game changer to drive operational cost efficiencies.

3. Improved Efficiency:

A predictive maintenance serves to assure that the equipment and machinery are operating at the most optimal performance levels. AI monitors the system health and performance continuously and can detect non radiant inefficiencies e.g. very small misalignments, friction or calibration problems, which can degrade output quality or energy efficiency. By addressing the issues before they affect performance, steady productivity, and reduction in energy use and production waste are maintained. This makes the operations more efficient, consumes fewer resources, and has overall throughput.

4. Extended Equipment Life:

However, all assets have a limited life but this can be increased with proper care and timely interventions. AI driven maintenance offers condition-based care where each piece of equipment is maintained as per actual usage and wear pattern rather than generic schedule. Using this approach will help to keep over maintenance to a minimum and prevent premature replacements so machinery and critical infrastructure will be kept down for far longer before replacement is required. This leads to a better ROI of capital asset and more sustainable practices of asset management.

Examples of AI in Predictive Maintenance:

- **Automotive:** The vehicles can be monitored by AI to know when maintenance is required such as tire pressure or wear in the brake pads.
- **Manufacturing:** AI can be used to monitor the performance of the machinery and be able to predict when the maintenance will have to be done, like the wear of a bearing, or vibration of the machine.
- **Energy:** The performance of power plants can be monitored by AI and the time when maintenance is essential, for example, turbine failure or generator overload, could be predicted.
- **Aviation:** Aircraft engine performance can be monitored and predicted as to when maintenance is required, such as engine wear or component failure, by AI.

5.4.2 How Machine Learning Helps Detect Anomalies

Anomaly detection is the detection of pattern or behaviour that is outside (not an expression of) or significantly deviates from the norm, which is often expressed as a bell curve. Typically, traditional rule base systems fail in dynamic and complex environment. This is where Machine Learning (ML) enters in taking a scalable, adaptive, intelligent way of spotting anomalies in real time.

Data related task of anomaly detection is an attempt to find outliers by algorithms. The outliers are important because the data could be inconsistent and these could be the cause of concern. It is important to identify outliers or anomalies in order to avoid misleading a data set or solving problems they may represent. Businesses can minimize their costs, retain customers and manage time when successful anomaly detection is in place.

Anomaly detection in machine learning

Machine learning models are used to detect anomalies rapidly in machine learning called anomaly detection. This serves several purposes, such as maintaining clean, high-quality data for processing or meeting specific business objectives. By ensuring quality data, organizations can have trust in their analysis, leading to better decision-making.

Although anomaly detection techniques have previously existed, more modern efforts that utilize machine learning can automatically detect outliers. The main advantages of anomaly detection in machine learning include the ability to handle significant volumes of data, high dimensional data from various sources, a high success rate in identifying anomalies, and the ability to have real-time detection.

1. Learning Normal Behaviour

ML models start by learning what "normal" looks like based on historical or live data. This could include system metrics like CPU usage, temperature, network activity, transaction volumes, or user behaviour. By processing large datasets, machine learning identifies the underlying patterns, trends, and relationships that define normal operations.

- **Supervised Learning:** Involves training the model with labelled data, where both normal and anomalous cases are known.

- **Unsupervised Learning:** No labelled data is required; the model identifies deviations from the majority of "normal" data on its own.
- **Semi-Supervised Learning:** Primarily trained on normal data, with occasional known anomalies to guide the model.

2. Real-Time Monitoring and Pattern Recognition

Once trained, ML models can monitor live data streams to compare current behaviour against learned patterns. Any significant deviation—such as a sudden spike in CPU usage, unusual login attempts, or out-of-sequence financial transactions—is flagged as an anomaly.

For example:

- A machine learning model monitoring a web server may detect a DDoS attack by noticing an unusual surge in requests.
- In manufacturing, ML may catch unexpected vibration patterns in a motor before it fails.

3. Detecting Subtle and Evolving Threats

In contrast to static rules, ML models can identify new or subtle irregularities that might not have been noticed previously. This is particularly helpful in identifying:

- Zero-day cyber attacks
- Slow-developing hardware failures
- Fraudulent financial activities
- Behavioural shifts in users or systems

These models can adapt over time through continuous learning, improving their accuracy and reducing false positives.

4. Types of Algorithms Commonly Used

- **Clustering (e.g., K-Means):** It groups similar data points and flags the point as outliers.
- **Isolation Forests:** Randomly partition the data in order to efficiently isolate anomalies.

- **Autoencoders (Neural Networks):** Measure and reconstruct input data—large errors indicate anomalies.
- **Time Series Forecasting (ARIMA, LSTM):** If observed values do not match with the predictions, they are marked as anomalies and projected future values.

Benefits of Machine Learning for Anomaly Detection

1. Increased Accuracy

Among the strong advantages of ML for anomaly detection is its much better accuracy. The ML algorithms are good at finding the subtle and complex patterns that can be missed by the human analysts or simple rule-based system. They are able to detect nonlinear relationships, multivariate dependencies, and also context-aware deviations, which greatly decreases false positives and false negatives rates. Especially in cybersecurity, finance, and other such fields where an early anomaly detection could save a lot of trouble, or money.

2. Efficiency

Machine learning allows machine processing and analysis of massive datasets at a high speed. When handling those large-scale environments such as a cloud infrastructure or IoT network, traditional tools may perform poorly, however, ML models can watch the real time data streams continuously and autonomously. This means that the organization can detect the anomaly as soon as it happens and then respond quickly and prevent exceeding damage or downtime.

3. Adaptability

The ability of ML to adjust to changing data is one of its most powerful features. Static rules become obsolete as users' behaviour changes, or as system size grows. However, ML models can retrain ones on new data, using recent trends and feedback. That way anomaly detection systems stay effective even in such fast-moving environments as e-commerce platforms during peak seasons or networks under the influence of recently appearing attacks.

4. Scalability

Anomalies are inherently scalable to ML based detection systems. No matter if you are observing a handful of sensors or a multitude of data streams, these models can be scaled horizontally to such a degree to support ever growing workloads without hampering performance. It is an important consideration for a corporation using big data or microservices, or wherever there are huge clusters of systems and companies located in multiple geographies.

5. Automation and Reduced Human Dependency

ML can help free up operations or security teams from the burden of running the process of detecting anomalies. The teams are brought smarter alerts and insights about scenario instead of combing logs or dashboards by hand. ML models can over time also recommend responses, or even autonomously take corrective action, e.g. isolate suspicious activity, or restart failing components.

6. Continuous Improvement

Over time, results from machine learning models get better for reasons: they're feeding more data into them and are getting feedback around their predictions in how they have performed. This improves the ability of detection, creates a more resilient, intelligent system that evolves with the business or technical infrastructure it is supporting.

One of the areas where machine learning (ML) can prove to be very beneficial is the field of anomaly detection; for instance, they can achieve high accuracy in detecting complex patterns and deviations, process real time data in instantaneous manner, and learn how to adapt with changes in the data. Because ML are very scalable, they are ideal for the large, distributed infrastructures. It also reduces human dependency through automation of detection and response and improve in the future through feedback loops. That's why ML is necessary in building resilient and insightful monitoring systems.

5.4.3 Automating Incident Response with AI

Automated Incident Response (AIR) is the use of software, algorithms, intelligence and automation (usually operating with Artificial Intelligence (AI) and Machine Learning (ML)) for monitoring, detecting, analyzing and responding to security incidents, anomalies and breaches in systems without direct human involvement. Modern cybersecurity and IT operations rely on it to make the incident handling faster, more accurate and consistent.

Disruptions in a company's infrastructure and systems can cause a huge effect on employees, customers, the brand reputation and ultimately the bottom line of the organization. To avoid damage and keep the operation going, these issues need to be addressed by the organization swiftly and efficiently.

Traditional manual incident response methods rely on human intervention, and human speed of response as the initial point of support, potentially also diverting attention from other critical duties. This manual approach often results in sluggish responses and is susceptible to errors stemming from human factors.

The fundamental stages of automated incident response

- **Identification of potential security incidents:** Within a security ecosystem, effective triage and automated incident response systems serve mainly as triage tools. They are made to react to current security point products that continuously monitor and conduct surveillance in order to spot any irregularities that might hint to a breach. Atypical user behaviour, malicious network connections, and questionable file uploads into the system are a few examples of these abnormalities. In order to help the appropriate security staff respond later, the automatic incident response system serves as a front-end sensor, gathering copious amounts of data from network devices including firewalls, servers, and endpoints.

- **Examination and analysis of accumulated data:** When the monitoring tools identify a security event, more security measures are used to examine the vast amount of data in an effort to identify trends and use artificial intelligence (AI) and machine learning (ML) to determine the incident's severity. This aids in keeping the investigation's focus. It can provide more details on attack types, duration, and source elements with real-time analytics. This makes it possible to determine the incident's primary cause.
- **Containment of compromised systems:** If the identity of a security incident is confirmed, the first step is to stop the threat as quickly as possible. The purpose of this process is to stop the spread of the incident in the network and to avoid further damage. Containing the incident is swift; and by swift containment, the impact of the incident is limited, and the potential harm is reduced, and escalation of the incident is mitigated.
- **Remediation of affected systems:** When the threat is contained, remediation becomes the focus and efforts are put towards restoring the affected systems to a healthy state. In this regard, the mitigation entails going through the vulnerabilities exploited by the incident, and addressing them systematically to prevent such an occurrence in the future. This may include software patching, recovering from backup, rebuilding of compromised systems. Through thorough remediating the affected systems, organizations reclaim operational continuity and minimize the probability for recurrence of similar incidents.
- **Post-incident analysis:** After the security incident's containment and remediation, organizations need to conduct a thorough post incident analysis. In this phase there will be a wide review of the incident response process to establish strengths, weaknesses

and also areas demanding improvement. The purpose of the post incident analysis is to determine root cause, evaluate the effectiveness, the response and to document and report.

5.5 SRE in Multi-Cloud and Hybrid Cloud Environments

Currently, as organizations move to multi/cloud and hybrid cloud strategies to maximize flexibility, performance, optimization for cost, and resiliency of the systems, Site Reliability Engineering (SRE) becomes a core discipline to support the operational complexity of such environments. SRE approach offers structured methodologies and tools and cultural practices for system reliability, scalability and observability regardless of where workloads are running.

These public cloud providers include AWS, Azure, Google Cloud, or Oracle Cloud in a multi cloud environment; allowing organizations to avoid vendor lock, to get region specific advantages, and scale better across fault. Hybrid cloud achieves by combining public cloud resources with a private cloud or on-premise infrastructure as to meet a regulatory requirements or maintain control over some sensitive data.

These architectures both offer as well as present problems. As services spread out across various platforms, visibility across the end to end, keeping things up, and taking actions automatically become even more important and difficult.

- **Hybrid Cloud**

Hybrid cloud is a combination of on-premises infrastructure (private cloud) and public cloud services where businesses can smoothly transfer workloads from one environment to another as and when required on business needs, cost basis, or as per any other regulation. At the same time, this model features a unified, flexible and scalable IT infrastructure, including traditional, legacy systems and the modern applications.

- **Multi-Cloud**

Multi-cloud refers to the use of multiple public cloud service providers (such as AWS, Azure, Google Cloud, or Oracle Cloud) to run different components of an organization's workloads or services. Instead of depending on a single vendor, businesses strategically distribute resources across clouds based on performance, cost-efficiency, service features, and geographic availability.

Table 5.1: Comparison of Hybrid Cloud vs. Multi-Cloud Architectures.

Feature	Hybrid Cloud	Multi-Cloud
Cloud Type	Combines private and public clouds	Uses multiple public clouds
Purpose	Balance control and scalability	Flexibility and vendor diversification
Management	Requires tools to manage disparate environments	Can be complex due to multiple providers
Security	Greater control over private infrastructure	Distributes workloads across multiple providers
Cost Optimization	Enables cost efficiency by shifting workloads as needed	Potentially higher cost due to redundancy and integration
Compliance	Easier to meet regulatory requirements via private cloud	Requires compliance alignment across providers
Performance	Can optimize performance-sensitive workloads locally	Enables geo-distributed performance tuning
Integration	Requires tight integration between environments	Involves interoperability between different cloud platforms
Disaster Recovery	Can use public cloud for backup and recovery	Increases resilience through multiple redundant services

Feature	Hybrid Cloud	Multi-Cloud
Vendor Lock-in	May involve some dependency on one public provider	Reduces dependency on any single cloud provider
Use Cases	Ideal for sensitive data handling and gradual cloud adoption	Suitable for global apps, high availability, and failover

Source: (Self-generated)

In a world of organization transition towards hybrid and multi cloud, Site Reliability Engineering (SRE) is what is needed to tackle the operational complexity these yields. The hybrid cloud combines the private and the public environment, the multi cloud multiplies the public provider. Both architectures offer their own benefits and disadvantages, so these need to have their own robust reliability, observability and automation practices.

5.5.1 Managing Reliability in Multi-Cloud Deployments

Hybrid cloud is the partial use of on-premises infrastructure (private cloud) combined with public cloud services which offers organizations the ease of moving workloads between environments as per convenience of business or as per the need of cost or any regulation. This model offers a unified, sophisticated, and very scalable IT infrastructure for modern cloud native apps as well as legacy system (Bauer & Adams, 2012).

Fragmenting monitoring and observability tools across a multi cloud environment is one of the main challenges in managing reliability across a multi cloud environment. Ecosystems may vary among them with performance metrics, logs, and alerts, and it is hard to manage the overall system health and performance. When visibility is in silos, one can end up missing incidents, improperly attributing the cause, and staying in recovery longer than needed.

Another problem is there are no SLAs or reliability standards across the cloud vendors. In the case of provider specific issue, the applications need to be designed to gracefully degrade, failover or rebalance. Whether this variability is consistent or inconsistent, predicts how users will experience the application or how the application will behave. It is without a strategy.

For these complexities, organizations have to adopt a strategic and systematic approach, and in many cases, Site Reliability Engineering (SRE). However, the structured framework that SRE uses is based on Service Level Objectives (SLOs), Error Budgets, automated incident response, and continuous improvement cycles. These principles assist teams in making decisions about risk, deciding what engineering work is important and putting reliability as a shared responsibility between development and operations.

Additional, SRE of today uses AI and machine learning to positively predict anomalies, failures and kick off automated remediation workflow on them. With IAC and policy driven governance, these tools then can be paired to standardize operations and provide reliability requirements in all platforms.

In the end, reliable multi cloud management is an issue not only technical but also cultural. To deliver in the fast pace of constant change it relies on collaboration across teams, a total knowledge of service dependency, commitment to continuous learning and iteration. Executed wisely, this technique not only safeguards the risks of an uncoordinated cloud economy but also unlocks the extent of the resilient, strong, and upcoming infrastructure.

Key Strategies for Managing Reliability

1. Unified Observability Across Clouds

Multi cloud requires it, and it's essential with a central monitoring platform. In order to enable SRE teams, observability tools need to be implemented, which are capable of ingesting and correlating metrics, logs, traces from all cloud environments into a single pane of glass. With platforms such as Prometheus, Grafana, Datadog or Open Telemetry you are able to see where the problems are and to understand the holistic failure path / degradation of performance early on.

2. Standardized SLOs and Error Budgets

All services are defined with Service Level Objectives (SLOs) and Error Budgets to guarantee consistent reliability across all services, no matter the cloud provider. It helps determine whether a service is operating within acceptable limits and help decide rollout speed or whether to stop new deploys to address stability trouble.

3. Cross-Cloud Redundancy and Failover

Redundant architecture is implemented to make sure that the critical services will be available despite an outage of one cloud provider. This may include:

- Multi-region and multi-zone deployments within each cloud.
- Load balancers and traffic routing mechanisms that can shift workloads between providers.
- It also ensures that data is replicated and synchronized across cloud environment and provide failover seamlessly.

4. Automated Incident Response

In systems with complex systems spread across clouds, there's need for automated alerting alert processing and response mechanism. Therefore, AI driven tools can also detect anomalies and start the automated workflows such as spinning up backup services, alerting the right teams, etc. Or execute the pre-defined scripts to remit the issues.

5. Infrastructure as Code (IAC)

Organizations can define infrastructure definition and ensure the consistency of infrastructure management using IAC tools like

Terraform, Pulumi, AWS CloudFormation or others. This provides reliable provisioning, reduces human error, as well as fast recovery with reproducible infrastructure.

Managing reliability on the multi cloud is complex. This higher flexibility, resilience, and performance also comes with more challenges in observability, SLA consistency, and incident management when we do not have a single cloud. This SRE structure that defines the unified monitoring, standardized SLOs, automation, and Infrastructure as Code helps hope to keep a various cloud reliable.

5.5.2 Avoiding Cloud Vendor Lock-In

An important aspect to avoid cloud vendor lock in when using cloud services is. Vendor lock in happens when a cloud user relies heavily on one cloud provider, especially when modifying across providers or migrating to a third-party cloud provider is difficult and expensive. However, as a limit to flexibility, raises the risk of running operations poorly, and leads to higher future costs since pricing is non-negotiable, proprietary technology is used, and it is hard, or even impossible, to scale or change the service over time to meet evolving business needs. Organizations have to design their infrastructure and applications to be portable from the ground up to mitigate these risks.

Multi-cloud is one of the most effective ways to avoid lock in. Using services from many cloud providers, the organization spreads the workload of its work on different platforms and thus reduces the dependence on a single vendor. This allows for the selectivity of the best possible services for each unique use case, for example, AWS for computing power, Google Cloud for ML and Azure for enterprise integration. Resilience is promoted in multi-cloud environments since if one provider goes down, the other can keep the operations going and avoid service disruptions.

Multi cloud strategies along with open standards, containerization, are also able to further reduce organizations from vendor lock in, by using technologies such as containers (Docker) and orchestration platforms (Kubernetes), others are able to package and run applications in a consistent environment in various clouds. The abstraction layer that this provides makes sure that applications are not dependent on particular cloud provider proprietary infrastructure or services. Also, they can use tools and practices like Infrastructure as Code (IaC) that automate and standardize provisioning of infrastructure thus making workloads' move among providers without special rework and industrial complications. When these strategies are combined, it enables businesses to increase flexibility, lower risks and still be able to select the right cloud provider to fit changing needs.

5.5.3 Hybrid Cloud Strategies for Enterprise Reliability

The Hybrid Cloud strategy for the enterprises has grown popular as a way to maximize the reliability, scalability, and flexibility. Hybrid cloud unites on premises infrastructures with public cloud infrastructures and allows the organization to retain major applications and sensitive data in private cloud and enjoy scalability and cost advantages of public cloud platforms. It gives enterprises the control of mission critical systems while allowing them to scale workload on demand. With a hybrid cloud, the infrastructure is resilient and it can be adaptable to any type of operational needs to boost reliability across various environments (Sehgal et al., 2023).

Hence, it becomes pivotal to concentrate on seamless integration between private and public clouds for reliable execution of enterprise workloads with hybrid cloud. It involves the use of tools and technologies that facilitate enabling the transfer of data, application interoperability and management in a smooth way between environments. Hybrid cloud solutions that an enterprise architect can provide need to be redundant as well as have failover capabilities: They must be capable of handling the load in case the cloud provider hosting their environment encounters an issue. Key to ensuring its optimization, automation, and orchestration is the use of resources allocation efficiently with the right number of workloads distributed across public and private clouds based on factors of performance, security, and compliance.

Hybrid cloud strategies also necessitate very robust monitoring and security practices in order to have a reliable system. To be able to pre-emptively identify issues that will impact an enterprise's availability and render service unavailable, enterprises must deploy the action to deploy for Unity observability tools that furnish real time visibility across both of their private and public clouds. However, since data security policies must be consistent across both environments, data is protected and compliance requirements are met, both environments should be similar. Enterprises can get a highly reliable and scalable infrastructure which can serve the requirements of a dynamic business landscape by bringing in hybrid cloud strategies, such deployment, standardization, and implementation of automation, monitoring, and security.

Multiple Choice Questions (MCQs)

1. **What is the primary purpose of implementing error budgets in SRE practices?**
   A) To set a strict limit on innovation
   B) To eliminate all system failures
   C) To restrict the number of user deployments
   D) To balance reliability with the pace of innovation

2. **How do error budgets help teams manage risk?**
   A) By allowing a controlled level of failure to foster innovation
   B) By predicting server outages in advance
   C) By ensuring 100% system uptime
   D) By replacing monitoring systems entirely

3. **What is a key benefit of strong collaboration between Dev and Ops teams in an SRE model?**
   A) Increasing the speed of siloed development
   B) Isolating roles to prevent conflict
   C) Reducing blame and promoting shared responsibility
   D) Reducing communication to only critical issues

4. **Which strategy best supports a shared responsibility model in SRE?**
   A) Encouraging collective ownership of system reliability
   B) Holding monthly blame assessments
   C) Separating operations from product teams
   D) Assigning failures solely to developers

5. **What does psychological safety in a culture of reliability encourage?**
   A) Ignoring minor issues to reduce workload
   B) Punishing errors to improve focus
   C) Preventing feedback to reduce friction
   D) Open communication and learning from failures

6. **What is the goal of establishing an SRE mindset across an organization?**

   A) Replacing product managers with reliability engineers
   B) Reducing the need for operations personnel
   C) Embedding reliability as a core organizational value
   D) Automating every task without oversight

7. **In what way does AI-driven predictive maintenance enhance system reliability?**

   A) By predicting and addressing failures before they occur
   B) By restricting deployments to off-peak hours
   C) By creating static dashboards
   D) By automating user feedback collection

8. **How does machine learning contribute to anomaly detection in SRE?**

   A) By manually labeling all incidents
   B) By identifying patterns and deviations that traditional tools may miss
   C) By removing historical logs after each cycle
   D) By randomly scanning system logs

9. **What is a major challenge in managing reliability in multi-cloud deployments?**

   A) Simplified monitoring across platforms
   B) Reduced network latency
   C) Fragmented observability and inconsistent SLAs
   D) High-speed deployments

10. Why might an enterprise adopt a hybrid cloud strategy for reliability?

A) To balance control, security, and scalability

B) To avoid regulation and compliance

C) To limit infrastructure to a single provider

D) To reduce internal collaboration

Answers

1.	2.	3.	4.	5.	6.	7.	8.	9.	10.
D	A	C	A	D	C	A	B	C	A

Bibliography

Adolph, R. (2016). *Reliability Engineering Theory and Practice.* 1–23.

Aiyenitaju, K. (2024). *The Role of Automation in DevOps : A Study of Tools and Best Practices.*

Aljuhani, A. A. (2017). *A multiple-criteria approach to support complex decisions in extreme programming. Doctoral thesis.* 1–235.

Arora, R., Kumar, A., Soni, A., & Tiwari, A. (2024). *AI-Driven Self-Healing Cloud Systems : Enhancing Reliability and Reducing Downtime through Event-Driven Automation AI-Driven Self-Healing Cloud Systems : Enhancing Reliability and Reducing Downtime.* https://doi.org/10.20944/preprints202408.1860.v1

Arun Kejariwal, J. A. (2017). *The Art of Capacity Planning Scaling Web Resources in the Cloud.*

Baines, T., Clegg, B., & An, A. (2014). *Growth through Servitization : Drivers, Enablers, Processes and Impact Proceedings of the Spring Servitization Conference 12-14 May 2014 Edited by ON CENT Confe rence Par rtners. May*, 9-31,32-37,48-54,55-61,62-69,88-94,95.

Bajgorić, N., Turulja, L., Ibrahimović, S., & Alagić, A. (2020). *Enhancing Business Continuity and IT Capability.* Auerbach Publications. https://doi.org/10.1201/9781003106098

Bauer, E., & Adams, R. (2012). Reliability and Availability of Cloud Computing. In *Reliability and Availability of Cloud Computing.* https://doi.org/10.1002/9781118393994

Baughan, S. (2023). *How risk management can improve your business.* Plooto.

Betsy Beyer, Niall Richard Murphy, David K. Rensin, Kent Kawahara, S. T. (2018). *The Site Reliability Workbook Practical Ways to Implement SRE.*

Bigelow, S. J. (2021). *Conduct a blameless postmortem and focus on the problem.* Tech Target.

Birman, K. P. (2005). Reliable Distributed Systems. In *Reliable Distributed Systems*. Springer-Verlag. https://doi.org/10.1007/0-387-27601-7

Blanchard, D. (2021). *Practices.*

Blank-Edelman, D. N. (2024). *Becoming SRE First Steps Toward Reliability for You and Your Organization.*

Bulmer, W. E. (2024). *A Constitution for the Common Good Strengthening Scottish Democracy After 2014.*

Chakravarty, S., & Chiang, Q. (2018). *H123954 (1).pdf.crdownload.*

Chinamanagonda, S. (2019). *Automating Infrastructure with Infrastructure as Code (IaC). 8* (11), 2037–2045.

Christie, A. M. (2012). *Software Process Automation The Technology and Its Adoption.*

Cooper Bethea, Gráinne Sheerin, Jennifer Mace, and R. K. with G. L. and G. O. (2019). *Managing Load.*

Cortex. (2022). *SRE Fundamentals: Everything you need to know.* Cortex.

Coupland, M. (2021). *DevOps Adoption Strategies: Principles, Processes, Tools, and Trends Embracing DevOps Through Effective Culture, People, and Processes.*

Crowe, M. K. (1993). Engineering Systems. In *Systems Science.* https://doi.org/10.1007/978-1-4615-2862-3_5

Dekker, S. (2017). *The Field Guide to Understanding 'Human Error.'* CRC Press. https://doi.org/10.1201/9781317031833

Digital, H. (2024). *Artificial intelligence in cybersecurity : Enhancing threat detection and prevention mechanisms through machine learning and data analytics Dimple Patil.* 39–43.

Donaldson, L. J., Rockville, W., Sorra, J., Gray, L., Streagle, S., Famolaro, T., Yount, N., Behme, J., Surugue, J., Vulto, A., Health Service Executive, Rafter, N., Hickey, A., Conroy, R. M., Condell, S., O'Connor, P., Vaughan, D., Walsh, G., Williams, D. J., … Lalor, D. J. (2020). Critical Appraisal Checklist. *Expert Opinion on Pharmacotherapy.*

Duderstadt, J. J. (2010). Engineering for a changing world a roadmap to the future of American engineering practice, research, and education. *Holistic Engineering Education: Beyond Technology*, 17–35. https://doi.org/10.1007/978-1-4419-1393-7_3

Eagle, E. (2023). *Connecting the Dots: A bias aware blueprint.*

Eckerson, B. W. W. (2012). What Are Performance Dashboards. *Performance Dashboards*, 3–22. https://doi.org/10.1002/9781119199984.ch1

Efraim Turban, Carol Pollard, G. W. (2021). *Information Technology for Management Driving Digital Transformation to Increase Local and Global Performance, Growth and Sustainability.*

Evans, David.S., Hagiu Andrei, S. R. (2019). How software platforms drive innovation and transform industries. In *Climate Change 2013 - The Physical Science Basis* (Vol. 53, Issue 9).

FasterCapital. (2020). Automating Repetitive Tasks. *FasterCapital.*

Gunderson, J. (2021). *Building and Scaling Your SRE Team.* Pagerduty.

Highsmith, J. A. (2002). *Agile Software Development Ecosystems.*

III, R. K. M. (2018). *A Model-Based Framework for Analyzing Cloud Service Provider Trustworthiness and Predicting Cloud Service Level Agreement Performance by Robert K . Maeser III B . A . in Computer Science, June 1984 , University of Minnesota M . S . in Software Engineerin. June 1993.*

Imperva. (2021). *Fault Tolerance.* Imperva.

INFOSECTRAIN. (2022). *Disaster Recovery (DR) Strategies.* INFOSECTRAIN.

Inspirisys. (2023). *Mastering the Art of Observability An Advanced Guide.* Inspirisys.

Israel Koren, C. M. K. (2020). *Fault-Tolerant Systems.*

Jackson, S. (2009). *Architecting Resilient Systems Accident Avoidance and Survival and Recovery from Disruptions.*

Jafarnejad Ghomi, E., Masoud Rahmani, A., & Nasih Qader, N. (2017). Load-balancing algorithms in cloud computing: A survey. *Journal of Network and Computer Applications, 88* (December 2016), 50–71. https://doi.org/10.1016/j.jnca.2017.04.007

Jez Humble, Joanne Molesky, B. O. (2014). *Lean Enterprise How High Performance Organizations Innovate at Scale.*

Karl E. Weick, K. M. S. (2015). *Managing the Unexpected Sustained Performance in a Complex World.*

King, B. (2010). Server Load Balancing. In *Performance Assurance for IT Systems.* https://doi.org/10.4324/9780203334577_chapter_26

Kleppmann, M. (2017). *Designing Data-Intensive Applications The Big Ideas Behind Reliable, Scalable, and Maintainable Systems.*

Konrad Hugo Jarausch, K. A. H. (2021). *Quantitative Methods for Historians A Guide to Research, Data, and Statistics.*

Kottisi, S. (2023). *Striking the Balance: A Comprehensive Guide to Site Reliability Engineering (SRE) Best Practices.* Medium.

Kunal, G. (2023). *Automate and Scale: Infrastructure as Code (IaC) in Cloud Architecture.*

Learning, C., & Reserved, A. R. (2010). Database Systems (9th Edition). In *Management.*

Lee, A. J. (2016). Systems thinking. In *Advanced Sciences and Technologies for Security Applications.* https://doi.org/10.1007/978-3-319-30641-4_2

Liang, Q. (2021). Continuous Delivery 2.0. In *Continuous Delivery 2.0.* https://doi.org/10.1201/9781003221579-1

Little, H. (2020). *Outdoor Learning Environments Spaces for Exploration, Discovery and Risk-taking in the Early Years.*

Lydia Parziale, Guillaume Lasmayous, Manoj S Pattabhiraman, Karen Reed, J. X. Z. (2014). *End-to-End High Availability Solution for System Z from a Linux Perspective.*

Makarova, I., Shubenkova, K., Mavrin, V., Mukhametdinov, E., Boyko, A., Almetova, Z., & Shepelev, V. (2020). Features of Logistic Terminal Complexes Functioning in the Transition to the Circular Economy and Digitalization. In *Lecture Notes in Intelligent Transportation and Infrastructure.* https://doi.org/10.1007/978-3-030-11512-8_10

Mistry, P. (2024). *Infrastructure as Code: Best Practices, Tools, Benefits, and More.* RADIX.

Montani, S., & Anglano, C. (2008). Achieving self-healing in service delivery software systems by means of case-based reasoning. *Applied Intelligence, 28* (2), 139–152. https://doi.org/10.1007/s10489-007-0047-1

Nancy McCormack, C. C. (2013). *Managing Burnout in the Workplace A Guide for Information Professionals.*

Niall Richard Murphy, Chris Jones, J. P. (2016a). *Site Reliability Engineering How Google Runs Production Systems.*

Niall Richard Murphy, Chris Jones, J. P. (2016b). *Site Reliability Engineering How Google Runs Production Systems.*

Penguin, B. (2024). *How AI Development Companies are Transforming Industries.* Bot Penguin.

Penny W. Cloft, Michael N. Kennedy, B. M. K. (2018). *Success is Assured Satisfy Your Customers On Time and On Budget by Optimizing Decisions Collaboratively Using Reusable Visual Models.*

Pethuru Raj Chelliah, Shreyash Naithani, S. S. (2018). *Practical Site Reliability Engineering Automate the Process of Designing, Developing, and Delivering Highly Reliable Apps and Services with SRE.*

Philosophie, M. Der. (2015). *MASTERARBEIT / MASTER ' S THESIS.*

Poulton, N. (2023). *The Kubernetes Book 2025.*

Pramatarov, M. (2024). *What is Load Balancing?* ClouDns.

Publishing, C. (2025). *The Future of the Book in the Digital Age.*

Reason, J. (2016). *Managing the Risks of Organizational Accidents.* Routledge. https://doi.org/10.4324/9781315543543

Religions, T. W. (2009). Routledge Handbook of. In *Political Science.*

Ricky Smith, R. K. M. (2011). *Rules of Thumb for Maintenance and Reliability Engineers.*

Rohrbeck, R. (2010). *Corporate Foresight Towards a Maturity Model for the Future Orientation of a Firm.*

Ruggles, W. S., & Harrington, H. J. (2018). *Project Management for Performance Improvement Teams.* Productivity Press. https://doi.org/10.4324/9781315117683

Saurav Bhattacharya, M. K. (2024). *Enterprise Digital Reliability Building Security, Usability, and Digital Trust.*

Sehgal, N. K., Bhatt, P. C. P., & Acken, J. M. (2023). Cloud Computing with Security and Scalability. In *Cloud Computing with Security and Scalability.* https://doi.org/10.1007/978-3-031-07242-0

Sharma, A. (2022). *DevOps Institute.* Devops Institute.

Sharma, S. (2017). *The DevOps Adoption Playbook A Guide to Adopting DevOps in a Multi-Speed IT Enterprise.*

Sheffi, Y. (2016). The power of resilience: how the best companies manage the unexpected. *Choice Reviews Online*, *53* (07), 53-3131-53–3131. https://doi.org/10.5860/choice.195121

Shkuro, Y. (2019). *Mastering Distributed Tracing Analyzing Performance in Microservices and Complex Systems.*

Singpurwalla, N. D. (2007). Reliability and Risk: A Bayesian Perspective. In *Reliability and Risk: A Bayesian Perspective.* https://doi.org/10.1002/9780470060346

Sky, O. (2020). *Battle Tested: How We Built and Master Our CI/CD Pipeline.* Medium.

Smith, D. J. (2021). *Reliability, Maintainability and Risk Practical Methods for Engineers.*

Squadcast. (2023). *Incident Response Tools: Key Considerations & Best Practices.* Squadcast.

Sutton, R. I. (2002). *Weird Ideas That Work 11 1/2 Practices for Promoting, Managing, and Sustaining Innovation.*

Tamer, C., Kiley, M., Ashrafi, N., & Kuilboer, J.-P. (2013). RISKS AND BENEFITS OF BUSINESS INTELLIGENCE IN THE CLOUD. *University of Massachusetts Boston, Management Science and Information Systems.*

Team, G. L. (2025). *A beginner's guide to distributed tracing and how it can increase an application's performance.* Grafana Labs.

Thompson, E. C. (2018). *Cybersecurity Incident Response How to Contain, Eradicate, and Recover from Incidents.*

TQP. (2022). *KPIs | Key Performance Indicators | KPI Examples.* TQp.

Turner, L. R. (2021). *Ebook: Becoming Agile: Coaching Behavioural Change for Business Results.*

Ukis, V. (2022). *Establishing SRE Foundations A Step-by-Step Guide to Introducing Site Reliability Engineering in Software Delivery Organizations.*

Veritis. (2022). *SRE vs DevOps: Which Productivity Approach is Better?* Veritis.

Vertisystem. (2023). *10 Load Balancing Techniques: Mastering the Art of Distributed Computing.* Medium.

Vikas Grover, Ishu Verma, P. R. (2023). *Achieving Digital Transformation Using Hybrid Cloud Design Standardized Next-generation Applications for Any Infrastructure.*

Vincent Faggiano, John McNall, Vincent/McNall Faggiano (Joh), T. T. G. (2011). *Critical Incident Management A Complete Response Guide.*

Weinberg, G. M. (2001). *Meaningful Project Feedback. XI* (1).

Wilensky, H. L. (2015). *Organizational Intelligence Knowledge and Policy in Government and Industry.*

Wiley, D. parmenter. (2020). *Key perfomance indicator developing, implementing and using winning KPIs.* 381.

Winn, D. C. E. (2017). *Cloud Foundry: The Definitive Guide Develop, Deploy, and Scale.*

Wright, S. (2022). Leadership Strategies for Improving Employee Engagement. *(Doctoral Dissertation, Walden University).*

Zeydan, E., Mangues-Bafalluy, J., Baranda, J., Martínez, R., & Vettori, L. (2022). A Multi-criteria Decision Making Approach for Scaling and Placement of Virtual Network Functions. *Journal of Network and Systems Management, 30* (2), 32. https://doi.org/10.1007/s10922-022-09645-9

About the Authors

My name is **Gopikrishna Maddali,** and I am based out of North Carolina, USA. I am currently working with a leading tech company, bringing over 18 years of rich experience in Database Architecture, IT Infrastructure Management, and Site Reliability Engineering (SRE). My career has been distinguished by a strong track record of architecting and optimizing database systems, enhancing infrastructure resilience, and implementing automation strategies tailored to the needs of large-scale enterprises, particularly in the healthcare domain.

As an experienced SRE leader, I specialize in building and maintaining highly available, fault-tolerant systems that scale effectively across distributed environments. I have designed and led the implementation of observability platforms using tools like Prometheus and Grafana, automated incident response workflows, and established SLIs, SLOs, and error budgets to align technical performance with business expectations. My focus on reliability, performance tuning, and continuous improvement has driven significant reductions in downtime and operational overhead across critical services.

I am currently pursuing an advanced course in Artificial Intelligence and Machine Learning from the University of Texas at Austin, expanding my skill set to include modern data-driven solution design. I also hold a Master of Computer Applications from India.

My Name is **Swapnil J. Wage.** I have an experience 15 years of experience in the software industry. I've grown from a hands-on software engineer to a team lead and, eventually, a senior engineering manager. My career has given me the chance to work on a wide range of projects, from building robust backend systems and designing databases to leading teams responsible for delivering reliable, high-performance applications.

Throughout my journey, I've always been passionate about bridging the gap between development and operations. I've helped teams adopt best practices for monitoring, incident management, and automation, ensuring that our systems are not just functional but also resilient and easy to maintain. I believe in empowering engineers, fostering a culture of ownership, and encouraging continuous learning and improvement.

As a mentor and leader, I've guided teams through architectural decisions, process improvements, and the challenges of scaling both technology and people. I'm a strong advocate for automation over manual work, proactive problem-solving, and building a collaborative environment where everyone can thrive.

This book is a reflection of my experiences and lessons learned along the way. Whether you're an engineer looking to grow your skills, a team adopting new practices, or a leader aiming to build a stronger engineering culture, I hope you'll find practical insights and inspiration within these pages."

www.ingramcontent.com/pod-product-compliance
Ingram Content Group UK Ltd.
Pitfield, Milton Keynes, MK11 3LW, UK
UKHW062310290726
14090UKWH00018B/991